Lecture Notes in Mathematics

A collection of informal reports and seminars
Edited by A. Dold, Heidelberg and B. Eckmann, Zürich

Series: Mathematisches Institut der Universität Bonn
Adviser: F. Hirzebruch

291

Peter Orlik
University of Wisconsin, Madison, WI/USA

Seifert Manifolds

Springer-Verlag
Berlin · Heidelberg · New York 1972

AMS Subject Classifications (1970): Primary: 57-02, 55F55, 57A10, 57E15
Secondary: 14J15, 55A05, 57D85

ISBN 3-540-06014-6 Springer-Verlag Verlin · Heidelberg · New York
ISBN 0-387-06014-6 Springer-Verlag New York · Heidelberg · Berlin

Offsetdruck: Julius Beltz, Hemsbach/Bergstr.

to Artie

Introduction

These are notes for a lecture series given at the University of Oslo in 1971 - 1972. Although the manifolds of the title were constructed by Seifert [1] in 1933, considerable interest has been devoted to them recently. The principal aim here is to survey the new results and to emphasize the variety of areas and techniques involved.

The equivariant theory comprising the first four chapters was initiated by Raymond [1], who discovered that two classes of Seifert manifolds coincide with certain fixed point free 3-dimensional S^1-manifolds. Chapter 1 contains Raymond's classification of S^1-actions on 3-manifolds. Chapter 2 describes equivariant plumbing of D^2-bundles over 2-manifolds and identifies the boundary 3-manifolds. This is used in chapter 3 to resolve singularities of complex algebraic surfaces with C^*-action. The technique is to compute the Seifert invariants of a suitable neighborhood boundary of the singular point and use these to construct an equivariant resolution following Orlik-Wagreich [1,2]. The equivariant fixed point free cobordism classification of Seifert manifolds due to Ossa [1] is given in chapter 4.

The remaining chapters contain topological results. The homeomorphism classification by Orlik-Vogt-Zieschang [1] using fundamental groups is obtained in chapter 5. The known free actions of finite groups on S^3 are given in chapter 6 following Seifert-Threlfall [1]. In chapter 7 we determine which Seifert manifolds fiber over S^1. The important results of Waldhausen [1,2] are outlined in the last chapter together with a number of other topics that we could not discuss in detail in the frame of the lectures.

I would like to thank my friends Frank Raymond and Philip Wagreich for teaching me directly or through collaboration much of the contents of these notes; the mathematicians in Oslo in general and Per Holm and Jon Reed in particular for their hospitality; and Professor F. Hirzebruch for inviting me to Bonn and for recommending the publication of these notes. Thanks are also due to Artie for thorough proofreadnig and to Mrs. Møller for careful typing of the manuscript.

Oslo, April 1972.

Peter Orlik *

*) Supported by grants from the National Science Foundation, the University of Oslo and the University of Wisconsin.

Contents

1. Circle Actions on 3 - Manifolds

In this chapter we introduce the necessary preliminary material concerning the action of a compact Lie group on a smooth manifold. Some important standard results are stated without proof.

We then proceed to the equivariant classification of circle actions on closed, connected, smooth 3-manifolds following Raymond [1] and Orlik and Raymond [1]. This is done in terms of a weighted 2-manifold (the orbit space together with information about the orbit types). It may be summarized as follows: the closed, connected, smooth 3-manifold M with smooth S^1 action is determined up to equivariant diffeomorphism (preserving the orientation of the orbit space if it is orientable) by the following set of invariants

$$M = \{b; (\epsilon,g,h,t); (\alpha_1,\beta_1),\dots,(\alpha_r,\beta_r)\} .$$

Here $\epsilon = o$ if the orbit space is orientable, $\epsilon = n$ if not; g is its genus; f is the number of components of fixed points in M ; t is the number of components of orbits with isotropy group $\mathbb{Z}_2$ and slice representation equivalent to reflection about a diameter in D^2 ; the relatively prime pair of positive integers (α,β) determines the orbit type of an orbit with isotropy group $\mathbb{Z}_\alpha$; and b is an integer representing an obstruction class subject to the conditions that $b = 0$ if $f+t > 0$, $b \in \mathbb{Z}$ if $f+t = 0$ and $\epsilon = o$, $b \in \mathbb{Z}_2$ if $f+t = 0$ and $\epsilon = n$ and $b = 0$ if $f+t = 0$, $\epsilon = n$ and some $\alpha_j = 2$.

Manifolds with $f+t = 0$ belong to the classes $0,o$ and N,nI of Seifert [1] and together with the other Seifert manifolds (introduced in chapter 5) will be the main topic of these notes.

1.1. Manifolds and Groups

A topological space X is a set with certain subsets U_i distinguished by being called open. The collection of open sets $\mathcal{U}$ is required to satisfy the following conditions:

(i) the empty set $\emptyset \in \mathcal{U}$ and $X \in \mathcal{U}$,

(ii) if $U, V \in \mathcal{U}$ then $U \cap V \in \mathcal{U}$,

(iii) if $U_i \in \mathcal{U}$, $i \in I$ then $\bigcup_{i \in I} U_i \in \mathcal{U}$ for an arbitrary index set I.

If $x \in X$ then an open neighborhood of x is an element of $\mathcal{U}$ containing x. A basis for the topology of X is a subcollection of open sets, $\mathcal{B}$ so that each element of $\mathcal{U}$ is a union of elements of $\mathcal{B}$. X is a Hausdorff space if for arbitrary distinct points $x_1, x_2 \in X$ there are open neighborhoods U_1, U_2 so that $U_1 \cap U_2 = \emptyset$. An open cover of X is a collection $\{U_i\}_{i \in I}$ of open sets so that $\bigcup_{i \in I} U_i = X$. A Hausdorff space is compact if for every open covering there exists a finite subcollection $\{U_{i_1}, \ldots, U_{i_n}\}$ which is an open covering of X. A map $f: X \to Y$ between topological spaces is continuous if the inverse image of every open set is open. It is a homeomorphism if there exists a continuous map $g: Y \to X$ so that $g \circ f = id_X$, $f \circ g = id_Y$. A space X is a topological manifold of dimension n if it is a Hausdorff space with a countable basis and every point $x \in X$ has an open neighborhood U_x homeomorphic to an open subset of Euclidean n-space $\mathbb{R}^n$. This homeomorphism $\varphi: U_x \to \mathbb{R}^n$ is called a coordinate system at x . Two coordinate systems φ and ψ are C^∞ related if $\varphi \circ \psi^{-1}$ and $\psi \circ \varphi^{-1}$ are C^∞ functions whenever defined. A set of coordinate systems $\mathcal{C}$ is a smooth structure on the topological manifold X if

(i) X is covered by the domains of the coordinate systems in $\mathcal{C}$,

(ii) any two coordinate systems in $\mathcal{C}$ are C^∞ related,

(iii) $\mathcal{C}$ is maximal with respect to (i) and (ii).

X is a smooth manifold if it has a smooth structure. A map $f: X \to Y$ between smooth manifolds is called a smooth map if for every two coordinate systems φ on X and ψ on Y the function $\psi \circ f \circ \varphi^{-1}$ is of class C^∞. A structure (topology, manifold, smooth) on X and Y induces a corresponding structure on the cartesian product $X \times Y$.

A group G is a topological group if G is a topological space and the group operations

$$(g_1, g_2) \to g_1 g_2 \quad \text{and} \quad g \to g^{-1}$$

are continuous maps. The topological group G is a Lie group if G is a smooth manifold and the above maps are smooth. Well known examples are the general linear group $GL(n;\mathbb{R})$ of $n \times n$ real invertible matrices, the orthogonal group $O(n)$ of $n \times n$ real orthonormal matrices and the special orthogonal group $SO(n)$ of $n \times n$ real orthonormal matrices with determinant $+1$. Note that $GL(n;\mathbb{R})$ is an open submanifold of $\mathbb{R}^{n^2}$ while $O(n)$ and $SO(n)$ are compact manifolds. A subgroup of a topological group is called closed if the corresponding subset is closed in the space of the group, i.e. its complement is open.

1.2. G - Manifolds

Let G be a compact Lie group and M a smooth manifold. A smooth (left) action of G on M is a smooth map

$$G \times M \to M$$
$$(g,x) \to gx$$

satisfying

(i) $g_1(g_2x) = (g_1g_2)x$

(ii) $ex = x$, where $e \in G$ is the identity element.

M together with the G action is called a G-manifold. If M_1 and M_2 are G-manifolds then the map $\varphi: M_1 \to M_2$ is called equivariant provided for all $g \in G$ and $x \in M_1$ we have $g\varphi(x) = \varphi(gx)$. Given $x \in M$ the subgroup of G defined by $G_x = \{g \mid gx = x\}$ is called the isotropy or stability group at x . The action is effective if only e leaves every point fixed, i.e. if $gx = x$ for all $x \in M$ then $g = e$. The subset of M defined by $Gx = \{gx \mid g \in G\}$ is called the orbit of x . The collection of isotropy subgroups along Gx , $\{G_{gx} \mid g \in G\}$ is called the orbit type. It is the conjugacy class of G_x in G since $G_{gx} = gG_xg^{-1}$. Consider the equivalence classes of orbits, $x \sim y \iff \exists g \in G \ni: y = gx$. Let x^* denote the equivalence class of x and M^* the collection of equivalence classes, called the orbit space, $M^* = M/G$. Let $\pi : M \to M^*$ be the orbit map. Topologize M^* by the quotient topology: U is open in M^* if and only if $\pi^{-1}(U)$ is open in M .
Notice that M^* is not a manifold in general. An action is transitive if for any two points $x,y \in M$ $\exists g \in G \ni: y = gx$, so all of M is one orbit and the orbit space is a single point. A G-manifold with a transitive action is called a homogeneous

space. A particularly important example of a homogeneous space is obtained as follows: Let G be a compact Lie group and H a closed subgroup. The coset space of H, G/H admits a natural action of G by multiplication and the action is clearly transitive.

1.3. G - Vector Bundles

A fiber bundle $\xi = (E,B,F,p)$ consist of a total space E, base space B, map $p: E \to B$ called bundle projection, a fiber F, an open cover $\mathcal{U}$ and for each $U \in \mathcal{U}$ a homeomorphism

$$\varphi_U : U \times F \to p^{-1}(U)$$

so that the composition $p \circ \varphi_U$ is projection onto the first factor. The structure group G of a fiber bundle is a group of homeomorphisms containing the homeomorphisms $F \to p^{-1}(b)$ defined by $x \to \varphi\,(b,x)$, and their inverses, for every $b \in B$. It is assumed that G acts on the above homeomorphisms transitively on the right. A fiber bundle is principal if the fiber is a topological group G which is also the structure group of the bundle. A vector bundle is a fiber bundle with fiber a vector space and structure group the general linear group of that vector space. Thus a real vector bundle has fiber $\mathbb{R}^n$ and group $GL(n)$. Typical example of a vector bundle is the tangent bundle TM of a smooth manifold M^n. The fiber at $x \in M$, $TM_x = \mathbb{R}^n$ and the total space of the bundle, TM is a smooth manifold of dimension $2n$. A G-vector bundle is a G-manifold M and a vector bundle with total space E over M so that there is a G-action on E compatible with the bundle structure, i.e. the map from $E_x = p^{-1}(x)$ to E_{gx} is an isomorphism making the diagram below commutative.

$$\begin{array}{ccc} G \times E & \longrightarrow & E \\ \downarrow{\scriptstyle id \times p} & & \downarrow{\scriptstyle p} \\ G \times M & \longrightarrow & M \end{array} .$$

Typical example is the tangent bundle TM of a G-manifold M. The map from TM_x to TM_{gx} is given by the differential of the map $g : M \to M$ evaluated at x.

Given $x \in M$ the map $gG_x \to gx$ defines an equivariant embedding $G/G_x \to M$ with image Gx, the orbit of x. Thus we may identify the G-manifolds G/G_x and Gx. Next we shall see that the normal bundle of Gx in M is naturally a G-vector bundle.

Let $E \to G/H$ be a G-vector bundle with base the homogeneous space G/H. Let V denote the fiber at eH. Since $h \in H$ leaves eH invariant, it leaves V setwise fixed so V is an H-module. Consider the principal H bundle $G \to G/H$ and the associated V bundle $G \times_H V$ over G/H obtained from $G \times V$ by identifying $[g,v] = [gh,h^{-1}v]$. Let G act on $G \times_H V$ by $k \in G$ $k[g,v] = [kg,v]$. Since $V \subset E$ given $g \in G$, $v \in V$ we have $gv \in E$, thus we have a map $[g,v] \to gv$ consistent with the identification, resulting in a map

$$G \times_H V \longrightarrow E$$

which is clearly a G-vector bundle isomorphism. Thus a G vector bundle over G/H is determined by the H-module structure of the fiber at eH.

Returning to the case when $H = G_x$, the normal bundle at $x \in Gx$ has fiber $V_x = TM_x/(TGx)_x$. For each $g \in G_x$ the differential of $g : M \to M$ induces a linear map $V_x \to V_x$ providing a representation $G_x \to GL(V_x)$ called the <u>slice representation</u>.

Its importance is given by the following theorem.

1.4. Some Basic Results

Slice theorem. Some G-invariant open neighborhood of the zero section of $G \times_{G_x} V_x$ is equivariantly diffeomorphic to a G-invariant tubular neighborhood of the orbit Gx in M by the map $[g,v] \to gv$ so that the zero section G/G_x maps onto the orbit Gx .

A proof is given in Jänich [1].

This gives at $x \in M$ a slice S_x with the following properties:

(i) S_x is invariant under G_x ,

(ii) if $g \in G$, $y, y' \in S_x$ and $g(y) = y'$, then $g \in G_x$,

(iii) there exists a "cell neighborhood" C of G/G_x so that $C \times S_x$ is homeomorphic to a neighborhood of x . If $\Gamma : C \to G$ is a local cross section in G/G_x then the map $F : C \times S_x \to M$ defined by $F(x,s) = \Gamma(c)s$ is a homeomorphism of $C \times S_x$ onto an open set containing S_x in M . In the differentiable case we may choose S_x as a suitably small closed disk in V_x .

Another useful theorem from the general theory of transformation groups is the following

Principal Orbit Type Theorem. Let M be a G-manifold and assume that M/G is connected. Then there is an orbit type (H) so that the orbits of this type, $M_{(H)}$ form a dense subset of M and the smooth manifold $M_{(H)}/G$ is connected. The type (H) is called principal orbit type, an orbit is called a principal orbit and the bundle $M_{(H)} \to M_{(H)}/G$ is called the principal orbit bundle.

A proof is given in Jänich [1].

We shall also use the following result.

<u>Conjugate Subgroup Theorem.</u> <u>Let</u> G <u>be a compact Lie group acting on a manifold</u> M . <u>If</u> $x \in M$ <u>and</u> $U \subset G$ <u>is an open set containing</u> G_x <u>then for</u> y <u>sufficiently near to</u> x , $G_y \subset U$.

A proof is given in Montgomery-Zippin [1, p.215].

1.5. The Circle Group

We are particularly interested in the circle group $G = S^1$. Recall first that there are different ways of thinking of this group:

(i) $G = U(1) = \{z \in \mathbb{C} \ \ |z| = 1\}$, complex numbers of modulus 1;

(ii) $G = SO(2)$, 2×2 real orthonormal matrices of determinant $+1$;

(iii) $G \cong T^1 = \mathbb{R}/\mathbb{Z}$, reals modulo the integers. (When convenient we shall think of the equivalent form $\mathbb{R}/2\pi\mathbb{Z}$, i.e. elements of G will be angles φ where $0 \leq \varphi < 2\pi$.)

Explicit isomorphisms are easily constructed and we shall use these three forms of G interchangably and without further warning. The closed subgroups of S^1 are (e) , the cyclic groups $\mathbb{Z}_\alpha$ and S^1 and by the Conjugate Subgroup Theorem the principal orbit type of an S^1 action is (e) . The purpose of this chapter is to give an equivariant classification of closed, connected 3-dimensional S^1-manifolds. First consider some examples.

1) Let

$$S^3 = \{z_1, z_2 \in \mathbb{C}^2 \mid z_1\bar{z}_1 + z_2\bar{z}_2 = 1\}$$

and define an action of $U(1)$ by $t \in U(1)$

$$t(z_1, z_2) = (t^\nu z_1, t^\mu z_2) .$$

This action is effective when $(\mu,\nu) = 1$. The orbit $\{z_1 = 0, z_2\bar{z}_2 = 1\}$ has isotropy group $\mathbb{Z}_\mu$ and the orbit $\{z_2 = 0, z_1\bar{z}_1 = 1\}$ has isotropy group $\mathbb{Z}_\nu$. All other orbits are principal. We shall see later that fixed point free S^1 actions on S^3 are in one-to-one correspondance with the pairs (μ,ν).

2) Consider S^3 as above with the action

$$t(z_1,z_2) = (z_1,tz_2) .$$

The action has one circle of fixed points, $\{z_2 = 0, z_1\bar{z}_1 = 1\}$ and all other orbits are principal. We shall see that this is the only action on S^3 with fixed points.

3) Take any closed 2-manifold B and let $M = B \times S^1$. Define an action of S^1 to be trivial in the first factor and the usual one in the second. This gives a free S^1 action with orbit space B.

4) Let $V = D^2 \times S^1$ be a solid torus with S^1 action trivial in the first factor and standard in the second. The subgroup $\mathbb{Z}_2 \subset S^1$ operates on the boundary with the principal (antipodal) action. If we collapse each of the orbits on the boundary of V by this $\mathbb{Z}_2$ action we obtain a closed manifold N with S^1 action. There are only principal orbits (corresponding to the interior of V) and orbits with isotropy group $\mathbb{Z}_2$ (correspon-ding to the boundary of V) that are doubly covered by nearby principal orbits so that the local orientation is reversed. The orbit space of the action is a disk with principal orbits in the interior and orbits with isotropy group $\mathbb{Z}_2$ on the boundary. The manifold N is the non-trivial S^2 bundle over S^1 called the non-orientable handle.

Before investigating the orbits with non-trivial isotropy

groups let us recall the orientation conventions of Raymond [1] and Neumann [1]. Given an oriented manifold M, its boundary ∂M is given the orientation which followed by an inward normal coincides with the orientation of M. If M is an oriented S^1 manifold and M^* is an orientable manifold, then we orient M^* so that M^* followed by the natural orientation of the orbits gives the orientation of M.

1.6. Fixed Points

Assume that $G_x = S^1$ so x is a fixed point. The slice at x may be chosen as a sufficiently small closed 3-ball D^3 and the action of G_x is an orthogonal action of S^1 on D^3. This is equivalent to the rotation of D^3 about an axis through x. The orbit space of this action on D^3 is a closed 2-disk with x on the boundary. So fixed points lie on 1-dimensional submanifolds and, by compactness, circles. A sufficiently small tubular neighborhood of one component of fixed points is therefore a solid torus with only fixed points and principal orbits. If we parametrize such a solid torus $V = D^2 \times S^1$ by (r,γ,δ) $0 \leq r \leq 1$, $0 \leq \gamma,\delta < 2\pi$ and let S^1 act by addition of angles, $0 \leq \theta < 2\pi$, then the action is equivalent to

$$\theta(r,\gamma,\delta) = (r,\gamma+\theta,\delta) .$$

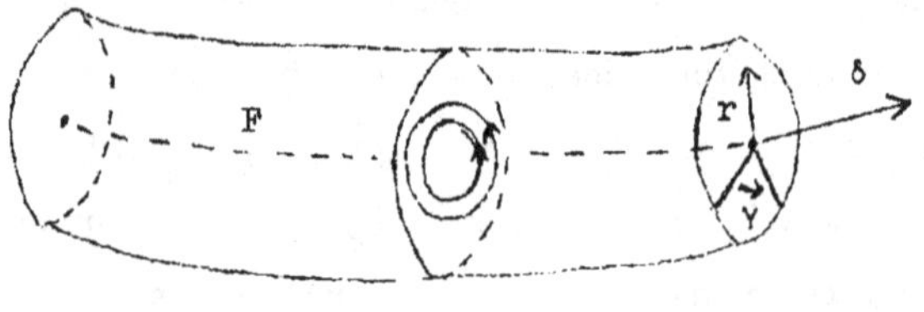

Call the collection of fixed points F and the (finite) number of components of fixed points f.

1.7. Exceptional Orbits

Let $G_x = \mathbb{Z}_\mu$. The orbit is 1-dimensional and the slice may be chosen as a 2-disk, D^2 . The actions of $\mathbb{Z}_\mu$ on D^2 are equivalent to rotation $(\mu > 2)$ and rotation or reflection $(\mu = 2)$. Consider the rotations in this section and the reflection in the next. Let $\xi = 2\pi/\mu$ act on the unit disk as follows

$$\xi(r,\gamma) = (r,\gamma + \nu\xi)$$

where $(\mu,\nu) = 1$ and $0 < \nu < \mu$.
We call this the <u>standard linear action</u> of type $[\mu,\nu]$. Since this is the action in each slice of such an exceptional orbit (called E-orbit), a small tubular neighborhood is a solid torus V with action equivalent to

$$\theta(r,\gamma,\delta) = (r,\gamma + \nu\theta,\ \delta + \mu\theta) \ .$$

The E-orbit corresponds to $r = 0$ and has isotropy group of order μ . We call $[\mu,\nu]$ the <u>oriented orbit invariants</u>. The corresponding <u>oriented Seifert invariants</u> (α,β) are defined by

$$\alpha = \mu \ , \quad \beta\nu \equiv 1 \mod \alpha \ , \quad 0 < \beta < \alpha \ .$$

Their geometric interpretation is the following.
Given an orientation on V, orient the slice so that it followed by the E-orbit gives the orientation of V . This orients the boundary of the slice m, a curve that is null-homotopic in V . Let l be a curve on ∂V homologous in V to the E-orbit and so that the ordered pair m,l gives the orientation on ∂V . Let h be an oriented principal orbit on ∂V . Since the action is principal on all of ∂V it admits a cross-section, q and any other section, q' is related to q by

$$q' = \pm\, q + sh$$

for some s . Orient q so that the ordered pair q,h gives the same orientation as m, l . Then we have

$$m = \alpha q + \beta h$$

and a suitable choice of s reduces β to the interval $0 < \beta < \alpha$.
Similarly

$$l = -\nu q - \rho h$$

for some ν and ρ so that

$$\begin{vmatrix} \alpha & \beta \\ -\nu & -\rho \end{vmatrix} = 1$$

thus $\beta\nu \equiv 1 \mod \alpha$.
Solving for q and h in the m, l system we have

$$q = -\rho m - \beta l$$
$$h = \nu m + \alpha l$$

Since l may be changed by $l' = l + sm$ we can reduce ν in the range $0 < \nu < \alpha$. In this case

$$\rho = (\beta\nu - 1)/\alpha .$$

In the action above, the curve

$$q = \{r = 1,\ \gamma = \rho\varphi,\ \delta = \beta\varphi,\ 0 \leq \varphi < 2\pi\} \subset \partial V$$

oriented by <u>decreasing</u> φ will satisfy the above conditions.

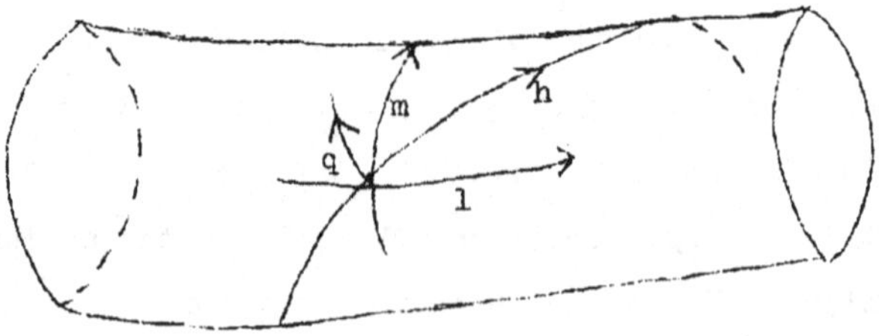

Changing the orientation on the solid torus V, keeping the action fixed results in a changed orientation for the slice and hence the slice inveriants change to $[\bar{\mu},\bar{\nu}] = [\mu,\mu-\nu]$. Similarly the Seifert invariants change to $(\bar{\alpha},\bar{\beta}) = (\alpha,\alpha-\beta)$. Thus the opposite orientation satisfies the condition

$$\beta\nu \equiv -1 \mod \alpha .$$

The latter was used in Orlik-Wagreich [1,2].

If there is no orientation specified on the solid torus V, then the orbit invariants are only defined as $[\mu,\nu]$, $0 < \nu \leq \mu/2$ and the Seifert invariants (α,β) , $0 < \beta \leq \alpha/2$ with $\nu\beta \equiv \pm 1$ mod α . We shall call these the <u>unoriented</u> orbit and Seifert invariants.

1.8. Special Exceptional Orbits

If $G_x = \mathbb{Z}_2$ and the action in the slice is reflection about an arc, then the neighborhood of such a special exceptional (SE) orbit is easily seen to be diffeomorphic to the cartesian product of the Moebius band with an interval, the non-trivial D^2 bundle over S^1 . All orbits intersecting the arc of reflection are SE-orbits, thus a component of SE-orbits is a torus. Let SE stand for the collection of SE-orbits and t denote the (clearly finite) number of components of SE .

1.9. The Orbit Space

As we have noted in the last three sections, the orbit space is a manifold near F^*, E^* and SE^* . It is clearly a manifold near principal orbits, so we conclude:

<u>Lemma</u> 1. <u>The orbit space</u> M^* <u>is a compact</u> 2-<u>manifold with boundary consisting of</u> $F^* \cup SE^*$.

Let us associate the symbol $\epsilon = o$ with an orientable and

$\varepsilon = n$ with a non-orientable orbit space and let g denote the genus in either case. If $\varepsilon = o$ we assume that an orientation of M^* is given. Thus we may associate the 4-tuple (ε, g, f, t) with M^* where $\varepsilon = o$ or n, $g \geq 0$, $f \geq 0$ is the number of boundary components in F^* and $t \geq 0$ is the number of boundary components in SE^*.

Lemma 2. *If $F \cup SE \neq \emptyset$ and $E = \emptyset$ then (ε, g, f, t) is a complete set of invariants for M up to equivariant diffeomorphism (preserving the orientation of M^* if $\varepsilon = o$).*

Proof. We show that the action admits a cross-section. Since $E = \emptyset$ we have a principal bundle over $M^* - F^* \cup SE^*$ and since $F^* \cup SE^* \neq \emptyset$ this bundle is trivial. Choose a cross-section to this bundle. It is now sufficient to extend this section in the neighborhood of each F-component and each SE-component. By (1.6) the neighborhood of an F-component is a solid torus V in M. The given cross-section restricted to ∂V is a torus knot of type $(1,b)$ for some b and it is well-known that there is an annulus in V spanned by this knot and the "center curve" (F-component) that extends the section. A similar argument applies to SE-components.

Next let us consider the somewhat more interesting case when $F \cup SE \cup E = \emptyset$. Here all orbits are principal and we have a bundle over the closed 2-manifold M^*. This bundle is classified by a map $M^* \to CP^\infty$ and hence by an element of $H^2(M^*;\mathbb{Z})$. This element is called the chern class or euler class of the bundle. If $\varepsilon = o$ then $H^2(M^*;\mathbb{Z}) = \mathbb{Z}$ and if $\varepsilon = n$ then $H^2(M^*;\mathbb{Z}) = \mathbb{Z}_2$ so the obstruction to the bundle being trivial is an integer b where $b \in \mathbb{Z}$ if $\varepsilon = o$ and $b \in \mathbb{Z}_2$ if $\varepsilon = n$.

We may interpret this integer b as follows: Remove the interior of a solid torus V_o from M . The remaining manifold, M_o admits a cross-section $\widetilde{M}_o^*$. Let q_o be the cross-setion to the action on the boundary oriented as the boundary of $-\widetilde{M}_o^*$. The equivariant sewing of the solid torus V_o into M_o is determined up to equivariant diffeomorphism by specifying the curve on the boundary of M_o

$$m = q_o + bh$$

that is to become nullhomotopic in V_o . We have obtained the following:

Lemma 3. *If* $E \cup F \cup SE = \emptyset$ *then* M *is determined up to equivariant diffeomorphism by* ϵ , g *and* b *where* $b \in \mathbb{Z}$ *if* $\epsilon = o$ *and* $b \in \mathbb{Z}_2$ *if* $\epsilon = n$.

In case $\epsilon = o$ the total space M is orientable. A change of orientation of M results in a change of sign for b .

We now have all the ingredients for the general case.

1.10. The Classification Theorem

Let S^1 *act effectively and smoothly on a closed, connected smooth* 3-*manifold* M . *Then the following orbit invariants*

$$M = \{b;(\epsilon,g,f,t);\ (\alpha_1,\beta_1),\dots,(\alpha_r,\beta_r)\}$$

subject to the conditions

(i) $b = 0$ *if* $f + t > 0$,

$b \in \mathbb{Z}$ *if* $f + t = 0$ *and* $\epsilon = o$,

$b \in \mathbb{Z}_2$ *if* $f + t = 0$ *and* $\epsilon = n$,

$b = 0$ *if* $f + t = 0$, $\epsilon = n$ *and* $\alpha_j = 2$ *for some* j ;

(ii) $0 < \beta_j < \alpha_j$, $(\alpha_j, \beta_j) = 1$ <u>if</u> $\epsilon = o$,
$0 < \beta_j \leq \alpha_j/2$, $(\alpha_j, \beta_j) = 1$ <u>if</u> $\epsilon = n$;

<u>determine</u> M <u>up to an equivariant diffeomorphism</u> (<u>which preserves the orientation of</u> M^* <u>if</u> $\epsilon = o$).

<u>Proof</u>. Given the above set of invariants a standard action is constructed as follows: Remove from M^* $(r+1)$ disjoint open disks $D_o^*, \ldots, D_r^*$. If $F \cup SE = \emptyset$ then the remaining manifold is a trivial principal bundle over $M^* - \bigcup_{j=o}^{r} D_j^*$ and admits a cross-section. If $F^* \cup SE^* \neq \emptyset$, remove these boundary components of $M^* - \bigcup_{j=o}^{r} D_j^*$, construct a cross-section and extend it to $F^* \cup SE^*$ as in (1.9.2). Let M_r be the resulting manifold with $(r+1)$ boundary components and let $\widetilde{M}_r^*$ be the cross-section. Sew in neighborhoods V_i of E-orbits with Seifert-invariant (α_j, β_j) $j = 1, \ldots, r$ next. Let Q be a boundary component of M_r^* and $Q \times S^1$ the corresponding boundary component of M_r . Let $Q \times \{0\}$ be the cross-section. Now sew the solid torus V of (1.7) equivariantly onto this boundary by mapping orbits onto orbits and the cross-section q of V onto $Q \times \{0\}$. Parametrize $Q \times S^1$ by $\{\gamma, \delta\}$, where increasing γ orients Q as a boundary component of $\widetilde{M}_r^*$.
Define the equivariant map

$$F : Q \times S^1 \to \partial V$$

by

$$F(\gamma, \delta) = (\rho\gamma + \nu\delta,\ \beta\gamma + \alpha\delta) .$$

Notice that

$$\begin{vmatrix} \rho & \nu \\ \beta & \alpha \end{vmatrix} = -1$$

and therefore F is orientation reversing as required. The

oriented cross-section q of ∂V maps onto the oriented curve $-Q$.

The equivariant sewing is therefore specified by the following. Given the cross-section $\tilde{M}_r^*$ in M_r let $q_0, q_1, \ldots, q_r$ be cross-sectional curves in ∂M_r oriented <u>opposite</u> to the induced orientation as components of $\partial \tilde{M}_r^*$. The equivariant sewing of the solid torus V_j $j = 1, \ldots, r$ makes the curve $m_j = \alpha_j q_j + \beta_j h$ on the j-th component of ∂M null-homotopic in V_j .

If $\epsilon = o$ then the pair (α_j, β_j) is determined in the interval $0 < \beta_j < \alpha_j$ and if $\epsilon = n$ only $0 < \beta_j \leq \alpha_j/2$ since the local orientation may be reversed along a path in M^* . We now have a manifold M_o with one torus boundary and a cross-section q_o to the action. We sew the last solid torus V_o fibered trivially onto this boundary so that the surve $m_o = q_o + bh$ becomes null-homotopic in V_o . This gives a manifold M with the required action.

Conversely, given an action on M, we shall recover its orbit invariants as follows: Read off ϵ, g, f, t from the orbit space, M^* . The equivariant tubular neighborhoods of E-orbits are isolated. Each one is equivariantly diffeomorphic to a solid torus V as described in (1.7) and the action is determined by the Seifert invariants (α, β) , $0 < \beta < \alpha$. If $\epsilon = n$ we use an isotopy of the tubular neighborhood along a path reversing the orientation on V^* to reverse the orientation on V . This reduces β to $0 < \beta \leq \alpha/2$. These pairs are invariants of V up to equivariant (orientation preserving, resp. not) diffeomorphism, specifying cross-sections $q_1, \ldots, q_r$ on the boundaries. If $F \cup SE \neq \emptyset$ these cross-sections may be extended to a global cross-section. If $F \cup SE = \emptyset$ and $\epsilon = o$ we have an obstruction in

$$H^2(M^* - \text{int}(V_1^* \cup \ldots \cup V_r^*),\ \partial(V_1^* \cup \ldots \cup V_r^*); \mathbb{Z}).$$

Its class is identified with the integer b. If $F \cup SE = \emptyset$ and $\varepsilon = n$ the above group equals $\mathbb{Z}_2$ and b may take on the values 0 or 1. A special argument shows that in the presence of an E-orbit of type (2,1) the two actions are equivariantly diffeomorphic, see Seifert [1, Hilfsatz VII].

It is easy to check that if M is orientable ($\varepsilon = o$ and $t = 0$), then a change of orientation results in the new orbit invariants

$$-M = \{b'; (o,g,f,0); (\alpha_1, \alpha_1 - \beta_1), \ldots, (\alpha_r, \alpha_r - \beta_r)\}$$

where $b' = 0$ if $f > 0$ and $b' = -b - r$ if $f = 0$.

In order to facilitate the notation we shall not insist that the Seifert invariants always be <u>normalized</u>. Writing M with these invariants should cause no confusion since the normalization is a well defined process.

Another notational convention will be the occasional use of the <u>orbit invariants</u> $[\mu, \nu]$ instead of the associated <u>Seifert invariants</u> (α, β). Again, the conversion is unique.

1.11. Remarks

1. The equivariant classification of (1.10) does not answer the question of when two S^1-manifolds are homeomorphic i.e., what are the possible different actions on a given manifold (c.f. the examples in 1.5). We shall call this the "topological classification problem".

(i) If $F \cup SE = \emptyset$ the manifolds involved coincide with Seifert's classes O,o and N,nI. These (together with the other Seifert manifolds introduced in chapter 5) are the central objects of our considerations and their mutual homeomorphism relationship will be discussed in detail in chapters 5 and 7 . These manifolds are irreducible with universal cover S^3 or R^3 .

(ii) If $F \neq \emptyset$ then the identification of the manifolds is done using equivariant connected sums. An arc S^* in the orbit space with end points on fixed point components and interior points corresponding to principal orbits has as inverse image under the orbit map a 2-sphere, S . Using such arcs the manifold is decomposed as the equivariant connected sum of 3-manifolds with the following orbit spaces.

$L^* =$ 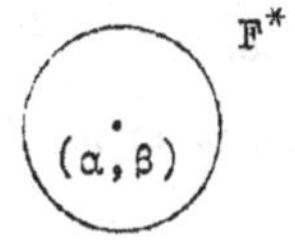

$L = \{0;(0,0,1,0);(\alpha,\beta)\}$

Clearly L is the result of an equivariant sewing of a solid torus neighborhood of F , V_1 and a solid torus neighborhood of the E-orbit, V_2 . Let h_i and q_i be the orbit and cross-section in ∂V_i . Then we have the relations for the bounding curves $m_1 = h_1$, $m_2 = \alpha q_2 + \beta h_2$. The equivariant sewing is $h_2 \to h_1$, $q_2 \to -q_1$ and going through the computations of (1.7) shows that we obtain the lens space $L(\alpha,\beta)$.

$M^* =$ 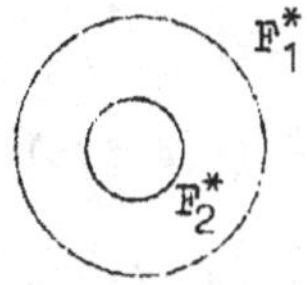

$M = \{0;(0,0,2,0\}$

Obviously $M = S^2 \times S^1$ with the standard S^1 action on the first factor and trivial action on the second factor.

$P^* =$ 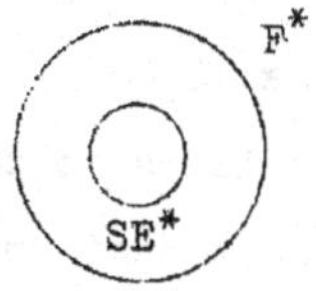

$P = \{0;(0,0,1,1)\}$

Similarly $P = P^2 \times S^1$ with the standard S^1 action on P^2 and trivial action on the second factor.

$N^* =$

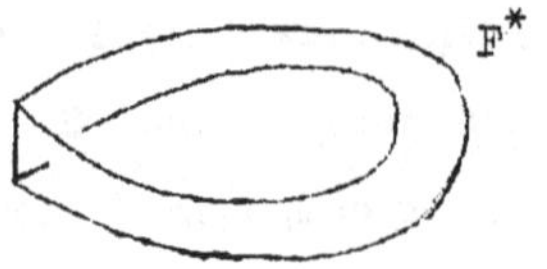

$N = \{0;(n,1,1,0)\}$

The manifold N is the non-orientable S^2 bundle over S^1. The action is visualized by taking $S^2 \times I$ with the usual S^1 action in the first factor and identifying $S^2 \times 0$ and $S^2 \times 1$ so that the orbits are reflected about the equator of S^2.

We state the following result without proof, Raymond [1].

<u>Theorem.</u> <u>Let</u>

$$M = \{b;(\varepsilon,g,f,t);\ (\alpha_1,\beta_1),\ldots,(\alpha_r,\beta_r)\}$$

<u>and assume that</u> $f > 0$. <u>Then</u> M <u>is equivariantly diffeomorphic to the equivariant connected sum:</u>

(a) $S^3 \# (S^2 \times S^1)_1 \# \ldots \# (S^2 \times S^1)_{2g+f-1} \# (P^2 \times S^1)_1 \# \ldots \# (P^2 \times S^1)_t$ $\# L(\alpha_1,\beta_1) \# \ldots \# L(\alpha_r,\beta_r)$ <u>if</u> $(\varepsilon,g,f,t) = (o,g,f,t)$, $t \geq 0$;

(b) $(S^2 \times S^1)_1 \# \ldots \# (S^2 \times S^1)_{g+f-1} \# (P^2 \times S^1)_1 \# \ldots \# (P^2 \times S^1)_t$ $\# L(\alpha_1,\beta_1) \# \ldots \# L(\alpha_r,\beta_r)$ <u>if</u> $(\varepsilon,g,f,t) = (n,g,f,t)$, $t > 0$;

(c) $N \# (S^2 \times S^1)_1 \# \ldots \# (S^2 \times S^1)_{g+f-2} \# L(\alpha_1,\beta_1) \# \ldots \#$ $L(\alpha_r,\beta_r)$ <u>if</u> $(\varepsilon,g,f,t) = (n,g,f,0)$.

(iii) The case $F = \emptyset$, $SE \neq \emptyset$ is handled using the classification of Seifert manifolds. The action lifts to the orientable double cover and commutes with the covering transformation. For details see Orlik-Raymond [1].

2. We assume that M is a smooth manifold and S^1 acts smoothly. It is known that all 3-manifolds are smoothable and using somewhat more elaborate arguments all the results hold for continuous S^1 actions on topological 3-manifolds, Raymond [1]. It follows from the discussion above that for the class of 3- manifolds with S^1 action the Poincaré conjecture holds.

3. Raymond [1] also studies the case when M is not compact. Allowing boundary makes the equivariant classification more cumbersome but essentially the same.

4. The classification above provides us with examples of manifolds that admit <u>no</u> S^1 <u>action</u> at all, e.g. any connected sum not on the list of the theorem.

2. Equivariant Plumbing

Plumbing is introduced for building blocks that are D^2 bundles over closed, orientable 2-manifolds, where it essentially consists of removing a $D^2 \times D^2$ from each of the objects and identifying the resulting boundaries after an interchange of factors. Prescribing an action of S^1 on the building blocks we may require that the plumbing respect this action. The resulting 4-manifold with boundary is studied in terms of the graph of the plumbing. The boundary is a closed, orientable 3-manifold with S^1 action and may be identified in terms of (1.10).

These ideas were first introduced by Hirzebruch [1] and von Randow [1]. The equivariant analogue was needed in Orlik and Wagreich [1] to resolve singularities of algebraic surfaces with C* action. This application is presented in the next chapter.

The orientation convention adopted here is that of Raymond [1]. The opposite was used in Orlik-Wagreich [1,2], where the letter b is also used differently.

2.1. Plumbing

The principal SO(2) bundles over a closed, orientable 2-manifold M are classified by $H^2(M;\mathbb{Z}) = \mathbb{Z}$. Denote the associated D^2 bundles indexed by $m \in \mathbb{Z}$ as $\eta = (Y_m, \pi, M)$. The compact 4-manifold Y_m has the homotopy type of M and if we let the zero section $\nu : M \to Y_m$ represent the positive generator $g \in H_2(Y_m;\mathbb{Z})$, then its self-intersection number $g \cdot g = m$ is the Euler class of Y_m. It is customary to let the bundle with Euler class $m = -1$ over S^2, $\eta = (Y_{-1}, \pi, S^2)$, be the disk bundle whose boundary, S^3, has the Hopf fibration.

Given two such bundles $\eta_1 = (Y_{m_1}, \pi_1, M_1)$ and $\eta_2 = (Y_{m_2}, \pi_2, M_2)$ we plumb them together as follows. Choose 2-disks $B_1 \subset M_1$ and $B_2 \subset M_2$ and the bundles over them, ξ_1 and ξ_2. Since they are trivial bundles there are natural identifications $\mu_1 : D^2 \times D^2 \to \xi_1$, $\mu_2 : D^2 \times D^2 \to \xi_2$. Consider the reflection $t : D^2 \times D^2 \to D^2 \times D^2$, $t(x,y) = (y,x)$ and define the homeomorphism $f : \xi_1 \to \xi_2$ by $f = \mu_2 t \mu_1^{-1}$. Pasting η_1 and η_2 together along ξ_1 and ξ_2 by the map f is called plumbing. It yields a topological 4-manifold with corners that may be smoothed. The resulting smooth manifold is independent of the choices involved.

A graph Γ is a finite, 1-dimensional, connected simplicial complex. Let $A_0, \dots, A_n$ denote its vertices. A star is a contractible graph where at most one vertex, say A_0, is connected with more that two other vertices. If there is such a vertex, call it the center. A weighted graph is a graph where a non-negative integer g_i (the genus) and an integer m_i (the weight) is associated with each vertex A_i.

Given a weighted graph Γ we define a compact 4-manifold $P(\Gamma)$ as follows: For each vertex (A_i, g_i, m_i) take the D^2 bundle $\eta_i = (Y_{m_i}, \pi_i, M_i)$ where M_i is a closed, orientable 2-manifold of genus g_i. If an edge connects A_i and A_j in Γ then perform plumbing on η_i and η_j. If A_i is connected with more then one other vertex, choose pairwise disjoint disks on M_i to perform the plumbing. Finally smooth the resulting manifold to obtain $P(\Gamma)$.

2.2. Equivariant Plumbing

We shall now define S^1 actions on the building blocks $\eta = (Y_m, \pi, M)$. For $g > 0$ let S^1 act trivially in the base and by rotation in each fiber. For $g = 0$ we define actions on $\eta = (Y_m, \pi, S^2)$ as follows: Let $S^2 = B_1 \cup B_2$ be the union of two 2-disks and $Y_m = B_1 \times D_1 \cup B_2 \times D_2$. Parametrize $D^2 \times D^2$ in polar coordinates with radii r and s , $0 \leq r, s \leq 1$ and angles γ, δ, $0 \leq \gamma, \delta < 2\pi$. The actions of S^1 on D^2 are equivalent to linear actions and we shall think of them as addition of angles. Let $\theta \in S^1$, $0 \leq \theta < 2\pi$. Define for $i = 1,2$

$$\theta_i : D^2 \times D^2 \to D^2 \times D^2$$

$$\theta_i(r, \gamma, s, \delta) = (r, \gamma + u_i\theta, s, \delta + v_i\theta)$$

Now Y_m is obtained by an equivariant sewing

$$G: \partial B_1 \times D_1 \to \partial B_2 \times D_2 .$$

Since the action is linear, G is determined by

$$F: \partial B_1 \times \partial D_1 \to \partial B_2 \times D_2$$

which in turn is isotopic to a linear map of the torus. Let F be defined by

$$F(\gamma) = x\gamma + y\delta \ , \ F(\delta) = z\gamma + t\delta .$$

Then F is equivariant if

$$u_1 x + v_1 y = u_2 \quad \text{and} \quad u_1 z + v_1 t = v_2 .$$

In order that G be equivariant on $\partial B_1 \times 0 \to \partial B_2 \times 0$ we need in addition that $u_1 x = u_2$, thus $y = 0$.

Since the determinant of F is -1 and the sewing results in a total space with euler class m, we need $x = -1$, $t = 1$, $z = -m$. Thus $u_2 = -u_1$, $v_2 = -mu_1 + v_1$. The action is effective if and only if $(u_1, v_1) = 1$.

A plumbing is <u>equivariant</u> if the identifying and trivializing maps are equivariant. Given a weighted graph Γ we say that

$P(\Gamma)$ is equivariant if each plumbing involved is equivariant. In that case the boundary $K(\Gamma) = \partial P(\Gamma)$ is a 3-manifold with S^1 action. We shall identify this manifold for certain graphs.

For $M = S^2$ we may think of the classifying element m as a map $S^1 \to S^1$ of degree $-m$. As above, ∂Y_m is obtained as the equivariant union of two solid tori

$$\partial Y_b = B_1^2 \times S_1^1 \underset{F}{\cup} B_2^2 \times S_2^1$$

where F has the matrix

$$\begin{pmatrix} -1 & 0 \\ -m & 1 \end{pmatrix} .$$

This is the sewing of two solid tori that results in the lens space $L(-m,1)$. Due to the well known diffeomorphisms $L(p,q) = - L(-p,q) = - L(p,p-q)$, we may write

$$\partial Y_m = L(-m,1) = L(m,m-1) .$$

Note also that the different actions on $L(-m,1)$ are given by the different pairs (u_1,v_1). For example $u_1 = 0$, $v_1 = 1$ $(u_2 = 0$, $v_2 = 1)$ gives the free action

$$L(-m,1) = \{-m;(o,0,0,0)\} .$$

In case $u_1 = 1$, $v_1 = 0$ we have a circle of fixed points and the orbit invariants are

$$L(-m,1) = \{0;(o,0,1,0);(m,m-1)\} .$$

Next consider the result of an equivariant plumbing according to the linear graph $\Gamma[b_1,\ldots,b_s]$

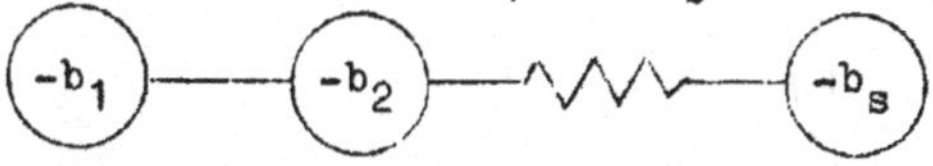

where each vertex has genus zero.

Lemma 1. The result of the equivariant linear plumbing according to the graph $\Gamma[b_1,\dots,b_s]$ above is the lens space $L(p_s,p'_s)$ where

$$\frac{p_s}{p'_s} = b_1 - \cfrac{1}{b_2 - \cfrac{1}{\ddots - \cfrac{1}{b_s}}} = [b_1,\dots,b_s]$$

Proof. Decompose each base space as $S_i = B_{i,1} \cup B_{i,2}$ with the corresponding trivializations of the bundles. As we have seen the first equivariant sewing requires $u_{1,2} = -u_{1,1}$ and $v_{1,2} = b_1 u_{1,1} + v_{1,1}$ so it has matrix

$$\begin{pmatrix} -1 & 0 \\ b_1 & 1 \end{pmatrix}.$$

Since the plumbing is equivariant the actions of $B_{1,2} \times S_{1,2}$ and $B_{2,1} \times S_{2,1}$ are the same but the factors are reversed, i.e. $u_{2,1} = v_{1,2}$ and $v_{2,1} = u_{1,2}$. The matrix of this map is

$$\begin{pmatrix} 0 & 1 \\ 1 & 0 \end{pmatrix}$$

and we have that

$$(u_{2,1}, v_{2,1}) = \begin{pmatrix} 0 & 1 \\ 1 & 0 \end{pmatrix} \begin{pmatrix} -1 & 0 \\ b_1 & 1 \end{pmatrix} \begin{pmatrix} u_{1,1} \\ v_{1,1} \end{pmatrix}.$$

The equivariant sewing of $B_{2,1} \times S_{2,1}$ and $B_{2,2} \times S_{2,2}$ has matrix

$$\begin{pmatrix} -1 & 0 \\ b_2 & 1 \end{pmatrix}$$

and the action on $B_{2,2} \times S_{2,2}$ is therefore expressed by

$$(u_{2,2}, v_{2,2}) = \begin{pmatrix} -1 & 0 \\ b_2 & 1 \end{pmatrix} \begin{pmatrix} 0 & 1 \\ 1 & 0 \end{pmatrix} \begin{pmatrix} -1 & 0 \\ b_1 & 1 \end{pmatrix} \begin{pmatrix} u_{1,1} \\ v_{1,1} \end{pmatrix} .$$

Continuing the sewing results in the equation

$$(u_{s,2}, v_{s,2}) = \begin{pmatrix} -1 & 0 \\ b_s & 1 \end{pmatrix} \begin{pmatrix} 0 & 1 \\ 1 & 0 \end{pmatrix} \begin{pmatrix} -1 & 0 \\ b_{s-1} & 1 \end{pmatrix} \cdots \begin{pmatrix} 0 & 1 \\ 1 & 0 \end{pmatrix} \begin{pmatrix} -1 & 0 \\ b_1 & 1 \end{pmatrix} \begin{pmatrix} u_{1,1} \\ v_{1,1} \end{pmatrix}$$

Note that all orbits are principal with the possible exception of the center curves of $B_{1,1} \times S_{1,1}$ and $B_{s,2} \times S_{s,2}$. The orbit space of the complement of these two solid tori is an annulus. Thus the total space is the result of the equivariant sewing of two solid tori by the product matrix above. Let

$$(u_{s,2}, v_{s,2}) = \begin{pmatrix} -p_{s-1} & -p'_{s-1} \\ p_s & p'_s \end{pmatrix} \begin{pmatrix} u_{1,1} \\ v_{1,1} \end{pmatrix} .$$

Then the total space equals the lens space $L(p_s, p'_s)$, where $p_s/p'_s = [b_1, b_2, \ldots, b_s]$. The latter fact follows from elementary properties of continued fractions, von Randow [1]. This completes the proof.

In particular if the action on $B_{1,1} \times S_{1,1}$ has an orbit of fixed points, $u_{1,1} = 1$, $v_{1,1} = 0$, then $B_{s,2} \times S_{s,2}$ has an E-orbit with <u>oriented orbit</u> invariants $[p_s, -p_{s-1}]$.

Next we shall show that equivariant plumbing imposes a strong condition on the shape of the graph provided the weights are negative. This will be the case for the applications in the next chapter.

<u>Lemma</u> 2. <u>Let</u> Γ <u>be a weighted graph and assume that</u> $P(\Gamma)$ <u>is equivariant.</u> <u>If</u>

(a) Γ has a vertex (A_0, g_0, m_0) where the action is trivial in the base,

(b) for each vertex (A_i, g_i, m_i) we have $m_i \leq -1$, and

(c) for each vertex $(A_i, 0, -1)$ connected with (A_j, g_j, m_j) we have $g_j > 0$ or $m_j \leq -2$ (or both) then

(i) $g_i = 0$ for all vertices $i > 0$,

(ii) Γ is a weighted star with center A_0,

(iii) the action is non-trivial on the base for $i > 0$.

Proof: Since we plumb around a fixed point, $0 \times 0 \subset D^2 \times D^2$, a vertex connected with more than two vertices must have trivial action in the base. Thus if A_1 is plumbed into A_0, it has non-trivial action in the base, hence $g_1 = 0$ and $u_{1,1} = 1$, $v_{1,1} = 0$. From above we get $u_{1,2} = -1$, $v_{1,2} = -m_1$. Define inductively $p_0 = 1$, $p_1 = -m_1$, $p_2 = -m_2 p_1 - p_0$, $p_j = -m_j p_{j-1} - p_{j-2}$, $j = 2, \ldots, r$. Then the action has $u_{j,2} = -p_{j-1}$, $v_{j,2} = p_j$. We define the auxiliary parameters $p_0' = 0$, $p_1' = 1$, $p_2' = -m_2$, $p_3' = -m_3 p_2' - p_1'$, $p_j' = -m_j p_{j-1}' - p_{j-2}'$, $j = 3, \ldots, r$. Then induction shows

1) $p_j p_{j-1}' - p_{j-1} p_j' = -1$ for $0 < j \leq r$,

2) $(p_j, p_j') = 1$, $(p_j, p_{j-1}) = 1$, $(p_j', p_{j-1}') = 1$ for $0 < j \leq r$,

3) if $-m_j \geq 1$ for $0 < j \leq r$ and if $-m_j = 1$ then $-m_{j\pm 1} > 1$ implies that we have $p_j \neq 0$ and $0 < p_j' < p_j$.

This proves the lemma.

Lemma 3. Consider the star S below with each $b_{i,j} \geq 2$ and $g_{i,j} = 0$ except for the center.

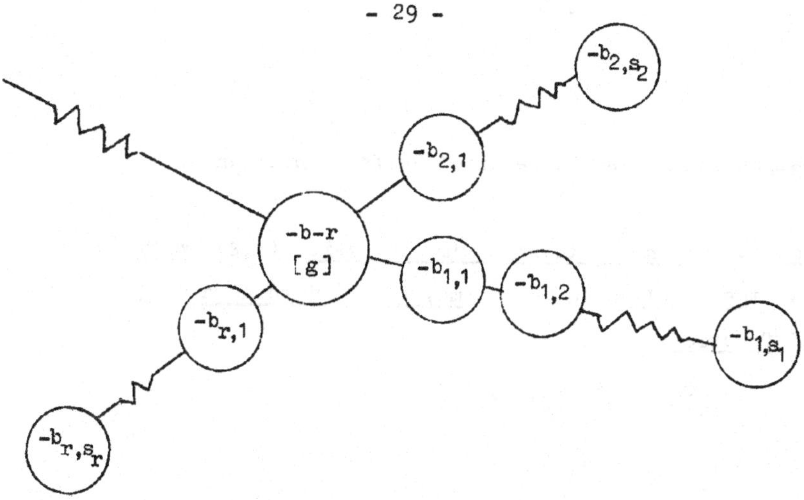

<u>The result of the equivariant boundary plumbing K(S) has Seifert invariants</u>

$$K(S) = \{b;(o,g,0,0);(\alpha_1,\beta_1),\ldots,(\alpha_r,\beta_r)\}$$

<u>where</u>

$$\frac{\alpha_j}{\alpha_j-\beta_j} = [b_{j,1},\ldots,b_{j,s_j}] \,, \qquad j = 1,\ldots,r \,.$$

<u>Proof</u>: By Lemma 1 each linear branch gives rise to a sewing of an E-orbit with orbit invariants $[p_{s_j}, -p_{s_j-1}]$. Since $p_{s_j} > 0$, $\alpha_j = p_{s_j}$ and $\nu = -p_{s_j-1}$.
From (1.7) and equation 1) above we have $\rho = p'_{s_j-1}$ and <u>before normalization</u> $\bar{\beta} = -p'_{s_j}$. According to 3) the <u>normalized</u> $\beta = \alpha + \bar{\beta} = \alpha - p'_{s_j}$. This proves the assertion that

$$\frac{p_{s_j}}{p'_{s_j}} = \frac{\alpha_j}{\alpha_j-\beta_j} = [b_{j,1},\ldots,b_{j,s_j}] \,.$$

The Seifert invariants of the manifold before <u>normalization</u> equal

$$K(S) = \{b+r;(o,g,0,0);(p_{s_1}, -p'_{s_1}),\ldots,(p_{s_r}, -p'_{s_r})\}$$

and normalization gives the required Seifert invariants.

Lemma 4. *Given relatively prime integers* (α,β) *with* $0 < \beta < \alpha$ *the fraction* $\alpha/\alpha-\beta$ *may be obtained as a unique continued fraction*

$$\frac{\alpha}{\alpha-\beta} = [b_1, b_2, \ldots, b_s]$$

where $b_i \geq 2$, $i = 1,\ldots,s$.

Proof: Repeated application of the Euclidean algorithm.

Corollary 5. *Every Seifert manifold*

$$K = \{b;(o,g,0,0);(\alpha_1,\beta_1),\ldots,(\alpha_r,\beta_r)\}$$

is the result of an equivariant plumbing according to a star $S(K)$ *as in Lemma 3.*

2.3. Quadratic Forms

Given a connected, oriented 4k-dimensional manifold M, a quadratic form S_M may be associated with it by homology intersections. Let $V = H_{2k}(M;\mathbb{Z})/$ torsion and define

$$S : V \times V \to \mathbb{Z}$$

by intersection of representative cycles. This is a well defined symmetric bilinear pairing, hence it induces a quadratic form on V , called S_M . As usual, the form may be diagonalized over the reals. Let p_+ denote the number of positive entries and p_- the number of negative entries. The integer

$$\tau(M) = \tau(S_M) = p_+ - p_-$$

is called the signature of the quadratic form (manifold). It is called positive (negative) definite if p_+ (p_-) equals the rank of V.

We want to compute the quadratic form of the compact 4-manifold $P(\Gamma)$. It is clear from the remarks of (2.1) that the graph Γ contains all necessary information. We may choose a basis for V consisting of one generator for each vertex (A,g,m) of Γ with self-intersection number m, and any two vertices connected in Γ have intersection number 1.

In particular the star corresponding to the Seifert manifold

$$K = \{b;(o,g,0,0);(\alpha_1,\beta_1),\dots,(\alpha_r,\beta_r)\}$$

$S(K)$ provided in (2.2.5) has quadratic form with matrix below where each unfilled entry equals zero.

$$S_M = \left[\begin{array}{c|cccc|cc|c|cc}
-b-r & 1 & & & & 1 & & & 1 & \\
\hline
1 & -b_{1,1} & 1 & & & & & & & \\
 & 1 & -b_{1,2} & \ddots & & & & & & \\
 & & \ddots & \ddots & 1 & & & & & \\
 & & & 1 & -b_{1,s_1} & & & & & \\
\hline
1 & & & & & -b_{2,1} & 1 & & & \\
 & & & & & 1 & -b_{2,s_2} & & & \\
\hline
 & & & & & & & \ddots & & \\
\hline
1 & & & & & & & & -b_{r,1} & 1 \\
 & & & & & & & & 1 & \ddots \\
 & & & & & & & & & 1 \quad -b_{r,s_r}
\end{array}\right]$$

Since $b_{i,j} \geq 2$ for all i,j this matrix is easily seen to be negative definite if and only if

$$-b-r < 0 .$$

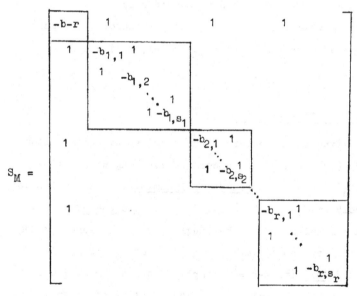

3. Resolution of Singularities

This chapter describes some results from Orlik-Wagreich [1,2]. Many of the ideas go back to Hirzebruch [1].

Given a complex algebraic surface with singularities, V admitting a "good" action of $\mathbb{C}^*$, the multiplicative group of complex members, we obtain a resolution of the singularities of V by the following method. If V has an isolated singularity, then a small neighborhood boundary S_ϵ invariant under the action of $U(1) \subset \mathbb{C}^*$ intersects it in $K = V \cap S_\epsilon$, a smooth, orientable, closed 3-manifold with S^1 action. Given the orbit invariants of K (1.10) we prove that the corresponding star (2.2.5) is the dual graph of a (canonical equivariant) resolution of the isolated singularity of V. If the singularity is not isolated then a normalization must preceed the above construction.

Rather than giving all the details as published, the emphasis here is on a survey of the background material, motivation and examples.

3.1. Algebraic and Analytic Sets

We shall define the necessary terminology as given in Fulton [1] and Gunning [1]. Let R be a commutative ring with unit. Let $R[X_1,\ldots,X_n]$ denote the ring of polynomials in n variables over R. A polynomial $F \in R[X_1,\ldots,X_n]$ is homogeneous of degree d if each monomial of F has degree d. An element $a \in R$ is irreducible if $a = b \cdot c$ implies that b or c is a unit. A ring R is a domain if $a \cdot b = 0$ implies $a = 0$ or $b = 0$. R is a UFD if every element has a unique factorization up to units and order. If R is a UFD so is $R[X]$. In particular $k[X_1,\ldots,X_n]$ is a UFD for any field k. The quotient field of $k[X_1,\dots,X_n]$ is the field of rational functions, $k(X_1,\dots,X_n)$. An ideal $I \subset R$ is proper if $I \neq R$, maximal if it is contained in no larger proper ideal and prime if $ab \in I$ implies either $a \in I$ or $b \in I$. An ideal is principal if it is generated by one element. A principal ideal domain (PID) is a domain where every ideal is principal. The residue classes of elements in R, modulo an ideal I, form a ring R/I and the natural map $\varphi : R \to R/I$ is a ring homomorphism. In particular $k[X_1,\dots,X_n]/I$ is a vector space over k. Given an ideal I, define its radical by $\operatorname{rad} I = \{a \in R \mid a^n \in I$ for some integer $n > 0\}$.

Let $\mathbb{C}^n$ be the affine complex n-space. If S is a set of polynomials in $\mathbb{C}[Z_1,\dots,Z_n]$ let $V(S) = \{\underline{Z} \in \mathbb{C}^n \mid F(\underline{Z}) = 0$ for all $F \in S\}$. Clearly $V(S) = \bigcap_{F \in S} V(F)$. A subset $X \in \mathbb{C}^n$ is algebraic if $X = V(S)$ for some S. Note the following properties:

(i) if I is the ideal in $\mathbb{C}[Z_1,\dots,Z_n]$ generated by S then $V(S) = V(I)$, so every algebraic set is equal to $V(I)$ for some ideal I;

(ii) if $\{I_\alpha\}$ is any collection of ideals, then $V(\bigcup_\alpha I_\alpha) = \bigcap_\alpha V(I_\alpha)$, so the intersection of any collection of algebraic sets is an algebraic set;

(iii) $V(F \cdot G) = V(F) \cup V(G)$, so any finite union of algebraic sets is an algebraic set;

(iv) if I defines an algebraic set then $I = \operatorname{rad} I$.

A ring is Noetherian if every ideal is finitely generated. In particular the Hilbert Basis Theorem shows that $\mathbb{C}[Z_1,\dots,Z_n]$ is Noetherian.

Projective complex n-space $\mathbb{CP}^n$ is defined as all lines through the origin $\underline{0} \in \mathbb{C}^{n+1}$. Any point $\underline{z} = (z_0,\dots,z_n) \neq \underline{0}$ defines a unique line $\{\lambda z_0,\dots,\lambda z_n \mid \lambda \in \mathbb{C}^*\}$ and two points $\underline{z}$, $\underline{z}'$

determine the same line if and only if there is a $\lambda \in \mathbb{C}^*$ so that $z_i = \lambda z_i'$ for all i. We let the equivalence class of these points $[z_0:z_1:\ldots:z_n]$ be the homeogeneous coordinates of a point in $\mathbb{CP}^n$. A projective algebraic set X is defined by homogeneous polynomials. It is irreducible if its ideal $I(X)$ is prime. In that case the residue ring $R_X = \mathbb{C}[X_0,\ldots,X_n]/I(X)$ is a domain called the homogeneous coordinate ring of X.

The ring of germs of holomorphic functions in n variables at $\underline{a} \in \mathbb{C}^n$ is denoted $\mathcal{O}_{\underline{a}}$. It is identified with the ring of convergent complex power series $\mathbb{C}\{z_1 - a_1,\ldots,z_n - a_n\}$. For $\underline{a} = \underline{0}$ call the ring simply $\mathcal{O}$. Note that for any two points $\underline{a}$, $\underline{b}$ the rings $\mathcal{O}_{\underline{a}}$ and $\mathcal{O}_{\underline{b}}$ are canonically isomorphic. The ring $\mathcal{O}$ is a Noetherian UFD. Its quotient field $\mathcal{M}$ is the field of germs of meromorphic functions at $\underline{0}$. The units of $\mathcal{O}$ are holomorphic germs not zero at $\underline{0}$. The ideal I of non-units in $\mathcal{O}$ is maximal and $\mathcal{O}$ is called a local ring. Note that $\mathcal{O}/I \approx \mathbb{C}$.

The sheaf of germs of holomorphic functions in n variables is also denoted $\mathcal{O}$. For any open set $U \subset \mathbb{C}^n$ there is a natural identification of the sections $\Gamma(U,\mathcal{O})$ with the ring $\mathcal{O}_U$ of holomorphic functions over U. For any point $\underline{a} \in \mathbb{C}^n$ the stalk of $\mathcal{O}$ at $\underline{a}$ is naturally the ring $\mathcal{O}_{\underline{a}}$ defined above.

An analytic sheaf $\mathcal{S}$ over an open set $U \subset \mathbb{C}^n$ is a sheaf of modules over the restriction $\mathcal{O}|U$. It is finitely generated over U if there are finitely many sections of $\mathcal{S}$ over U which generate the stalk $\mathcal{S}_{\underline{a}}$ as an $\mathcal{O}_{\underline{a}}$ module at each point $\underline{a} \in U$.

An analytic subvariety X of an open set $U \subset \mathbb{C}^n$ is a subset of U which in some open neighborhood of each point of U is the set of common zeros of a finite number of functions defined and holomorphic in that neighborhood. Two such pairs (X_1,U_1),

(X_2, U_2) are equivalent if there is an open neighborhood $W \subset U_1 \cap U_2$ so that $W \cap X_1 = W \cap X_2$. The equivalence class is called a <u>germ of an analytic subvariety</u>. The ideal of the subvariety at a point is defined for the origin by $I(X) = \{f \in \mathcal{O}_0 \mid \exists$ analytic subvariety X of $U \subset \mathbb{C}^n$ representing the germ X and an analytic function $f \in \mathcal{O}_U$ representing the grem f with $f|_X \equiv 0\}$. A germ X is said to be <u>reducible</u> at $\underline{a}$ if $X = X_1 \cup X_2$ where X_i are also germs of analytic subvarieties at $\underline{a}$; otherwise it is <u>irreducible</u> at $\underline{a}$.

An <u>analytic variety</u> is a Hausdorff space V with a distinguished subsheaf $\mathcal{O}_V$ of the sheaf of germs of continuous complex valued functions on V so that at each point $\underline{a} \in V$ the germ of V together with the stalk $(\mathcal{O}_V)_{\underline{a}}$ is called the <u>sheaf of germs of holomorphic functions</u> on V. A <u>morphism</u> between analytic varieties V and V' is a continuous mapping $\varphi: V \to V'$ so that $\varphi^*(\mathcal{O}_{V'}) \subset \mathcal{O}_V$. A point in an analytic variety V is <u>regular</u> (simple) if the germ of V at that point is equivalent to the germ of $\mathbb{C}^n$ for some n. The set of all regular points is the <u>regular locus</u> of V. It is an analytic manifold, not necessarily connected or pure dimensional. Its complement in V is called the <u>singular locus</u> and a point on it a <u>singular</u> point. The variety is called <u>non-singular</u> if the singular locus is empty. A singular point $\underline{x}$ is <u>isolated</u> if there is a germ at $\underline{x}$ with no other singular points.

Notice that if V is algebraic in $\mathbb{C}^n$ then $I(V)$ is finitely generated, say $I(V) = (g_1, \ldots, g_r)$. The Jacobian matrix

$$J(V) = \frac{\partial(g_1, \ldots, g_r)}{\partial(z_1, \ldots, z_m)}$$

has maximal rank, $\operatorname{rk} J(V) = m - n$ at regular points and at singular points $\operatorname{rk} J(V) < m - n$.

3.2. Intersections and Covers

Let V be a non-singular complex analytic surface. The algebraic intersection pairing

$$H^2(V) \otimes H^2(V) \to \mathbb{Z}$$

is defined using Poincaré duality

$$\Delta : H^2(V) \to H_2(V) .$$

For $X, Y \in H^2(V)$ define the pairing by

$$(X,Y) \to (X \cdot Y) = X(\Delta Y) .$$

Recall that in case V is not compact we use homology with closed supports in the definition of Δ .

A map $\varphi : V' \to V$ is said to be <u>proper</u> if the inverse image of a compact set is compact. If φ is a proper surjective map of analytic spaces of dimension n, then there is a positive integer d and an open subset $U \subset V$ so that $\varphi^{-1}(v)$ consists of d points for all $v \in U$. We call d the degree of φ . If V and V' are complex surfaces, φ is a map of degree d and D_1 and D_2 are elements of $H^2(V)$, then $(\varphi^*(D_1) \cdot \varphi^*(D_2)) = d(D_1 \cdot D_2)$.

Let X, X' be curves in a non-singular surface V and $x \in X \cap X'$. We say that X meets X' <u>normally</u> at x if there is a coordinate neighborhood U of x and local coordinates z_1 and z_2 so that $X \cap U$ is the locus $z_1 = 0$ and $X' \cap U$ is the locus $z_2 = 0$. It is well known that if $X \neq X'$ and $(X \cdot X') = 1$ then X meets X' normally at precisely one point.

We say that φ is a <u>finite</u> map if φ is proper and $\varphi^{-1}(v)$ consists of a finite number of points for all $v \in V$. Suppose moreover that φ is surjective. The set B of points $v \in V$, so that $\varphi^{-1}(v)$ consists of fewer than $d = \text{degree}\,\varphi$ points, is called the <u>branch locus of</u> φ . It is well known that if V is non-singular then B is the union of a finite number of irreducible subvarieties <u>each</u> of complex codimension 1 ("purity of the branch locus").

Suppose X is a curve on a surface V . If $x \in X$ we recall that X is <u>locally irreducible</u> at x if for every sufficiently small neighborhood U of x in V there is a unique irreducible component of $X \cap U$ containing x . If $x \in X$ then there is a neighborhood U of x in V so that $X \cap U = X_1 \cup \ldots \cup X_r$, where each X_i is a curve which is locally irreducible at x . The X_i are called <u>the branches of</u> X_i <u>through</u> x .

<u>Definition 1</u>. Suppose $\varphi : V' \to V$ is a finite map of non-singular surfaces or curves, B is the branch locus of φ and $\varphi(v') = v \in B$. Let X_i be a branch of $\varphi^{-1}(B)$ passing through v' (in the case of curves this is just v'). There is a neighborhood U of v in V and a holomorphic function f in U having a zero of order 1 along $B \cap U$ and no other zeros. Let $e(X_i)$ equal the order of the zero of $f \circ \varphi$ along X_i . This is called the <u>ramification index of</u> φ <u>along the branch</u> X_i at v' . Now

$$\sum_{\substack{v' \in \varphi^{-1}(v) \\ v' \in X_i}} e(X_i) = \text{degree } \varphi$$

where we let X_i range over all branches of $\varphi^{-1}(B)$ through v'. If there is a unique branch of $\varphi^{-1}(B)$ through v', we denote $e(X_i)$ by $e(v')$. In this case we get $\sum_{v' \in \varphi^{-1}(v)} e(v') = \text{degree } \varphi$. Note that $v \in B$ if and only if $e(v') > 1$ for some $v' \in \varphi^{-1}(v)$.

If X is an irreducible curve on a non-singular analytic surface V, then there is an open dense subset $Y \subset X$ with the property that X is locally irreducible at all points of Y .

Suppose $\varphi^{-1}(X) = X_1 \cup \ldots \cup X_r$ where the X_i are irreducible. Then there is an open dense subset Y' of X so that $Y' \subset Y$, $X_i \cap \varphi^{-1}(Y')$ is locally irreducible and for any $v_1, v_2 \in X_i \cap \varphi^{-1}(Y')$ we have $e(v_1) = e(v_2)$. Call this integer $e(X_i)$, the ramification index of X_i over X. It follows immediately from the definition of φ^* that

$$\varphi^*(X) = \sum_{i=1}^{r} e(X_i)X_i \in H^2(V') .$$

We can use the ramification index to get a useful relation between the genus of an analytic curve and the genus of a finite cover of that curve.

Proposition 2. (Hurwitz formula) <u>Let $\varphi : X' \to X$ be a finite morphism of compact non-singular complex curves. Let</u> $2g_X = \dim H^1(X, \mathbb{Z})$, $2g_{X'} = \dim H^1(X', \mathbb{Z})$. <u>Then</u>

$$(2 - 2g_{X'}) = (\text{degree } \varphi)(2 - 2g_X) - \sum_{x' \in X'} (e(x') - 1) .$$

Proof. Triangulate X so that the points of the branch locus are vertices of triangles and no two are connected by a 1-simplex. The Euler number of the triangulation is $2 - 2g_X$. It can be lifted to a triangulation of X' by means of φ since outside of B the map φ is a local homeomorphism. This multiplies the number of faces and edges by degree φ. If $x \in X$ is a vertex and $x \notin B$, then there are degree φ vertices above x. But if $x \in B$, then there are degree $\varphi - \sum_{\varphi(x')=x} (e(x') - 1)$ vertices above x. This proves the formula.

3.3. Monoidal Transforms and Resolution of Singularities

Definition 1. Suppose V is an analytic space, $\mathcal{O}_V$ is the sheaf of holomorphic functions on V and $I \subset \mathcal{O}_V$ is an ideal sheaf. The monoidal transform with center I is a pair (π, V') with $\pi : V' \to V$ and

(i) $I\mathcal{O}_{V'}$ is locally principal i.e. $\forall v \in V'$ the stalk $(I\mathcal{O}_{V'})_v$ is generated by one function,

(ii) for every $\pi_0 : V_0 \to V$ satisfying "$I\mathcal{O}_{V_0}$ is locally principal" there is a unique $\sigma : V_0 \to V'$ with $\pi \circ \sigma = \pi_0$.

The monoidal transform exists, Hironaka [1,p.129], and is unique by (ii). If X is a subspace of V and I_X is the sheaf of functions vanishing on X, then the monoidal transform with center X is just the monoidal transform with center I_X.

We can construct the monoidal transform as follows. Suppose $v \in V$. Then there is a neighborhood U of v and holomorphic functions $f_0,\ldots,f_r$ on U so that the restriction of I to U is generated by $f_0,\ldots,f_r$. Let X be the set of common zeros of the f_i. These functions define a map

$$\varphi : U - X \to \mathbb{CP}^r$$

by $\varphi(u) = [f_0(u):\ldots:f_r(u)]$. Let

$$\Gamma \subset (U - X) \times \mathbb{CP}^r$$

be the graph of φ, let V'_U be the closure of Γ in $U \times \mathbb{CP}^r$ and let

$$\pi_U : V'_U \to U$$

be the projection map. Then (π_U, V'_U) is the monoidal transform with center $I|U$. If we choose an open cover $\{U_i\}$ of V where the U_i are as above, then the universal property of monoidal transforms guarantees that the (π_{U_i}, V'_{U_i}) piece together to give (π, V'). Note that if Y is the set of common zeros of the functions in I, then $V - Y$ is an open dense subset of V and $\pi : \pi^{-1}(V - Y) \to V - Y$ is an isomorphism. The monoidal transform with center $\{v\}$ is also called the σ-transform with center at v.

<u>Definition 2</u>. Suppose V is an analytic space and $X \subset V$ is the set of singular points of V. We say that $\pi : V' \to V$ is a <u>resolution of the singularities of</u> V if

(1) π is proper,

(2) V' is non-singular,

(3) π induces an isomorphism between $V' - \pi^{-1}(X)$ and $V - X$.

<u>Remark</u>. It is known, Hironaka [1], that if V is an algebraic surface, then there is a resolution π which is a composite of monoidal transforms. For an isolated singularity we shall construct a "canonical" resolution but first we need a definition.

<u>Definition 3</u>. An analytic space V is said to be <u>normal at</u> $v \in V$ if for every neighborhood U of v and meromorphic function f on U and holomorphic functions $\{a_i\}$ on U, the equation

$$f^n + a_{n-1} f^{n-1} + \dots + a_0 = 0$$

implies that f is holomorphic. V is said to be normal if V is normal at every $v \in V$. A curve is normal if and only if it is non-singular. On a normal variety V the singular locus has codimension ≥ 2. If $v \in V$ is a simple point, then v is a normal point. For any analytic variety V there is a unique pair $(\pi, \tilde{V})$ so that $\pi : \tilde{V} \to V$, $\tilde{V}$ is normal and for any normal variety

V' and $\pi: V' \to V$ there is a unique map $\sigma: V' \to \tilde{V}$ with $\pi \circ \sigma = \pi'$. The pair $(\pi, \tilde{V})$ is called the <u>normalization</u> of V. The map π is finite and it is an isomorphism over an open dense subset of V.

Suppose V is a complex algebriac surface with an isolated singular point v. There is a finite sequence of maps $\pi_i: V_i \to V_{i-1}$, $i = 1, \ldots, n$ so that $V_0 = V$, V_n is non-singular; π_i is a normalization if i is even and π_i is the monoidal transform with center at the (isolated) singular points of V_{i-1}. Thus V_n is a resolution of $v \in V$ but $\pi^{-1}(V)$ may be rather complicated.

In order to improve $\pi^{-1}(v)$ we perform a further sequence of monoidal transformations $\pi_{n+j}: V_{n+j} \to V_{n+j-1}$ so that the composite $\pi = \pi_1 \ldots \pi_{n+k}$ satisfies

(*) $\pi^{-1}(v) = X_1 \cup \ldots \cup X_r$, the X_i are non-singular irreducible curves, $(X_i \cdot X_j) = 0$ or 1 for $i \neq j$ and $X_i \cap X_j \cap X_k = \emptyset$ for distinct i, j, k.

Let $\sigma_i = \pi_1 \circ \ldots \circ \pi_i$. Then we can choose π_{n+j} so that it is the monoidal transform with center $x \in V_{n+j-1}$ where either

(1) x is a singular point of some component of $\sigma^{-1}_{n+j-1}(v)$

(2) x is a point of $X_i \cap X_j$ and X_i and X_j do not meet normally at x,

(3) x is a point of $X_i \cap X_j$ and $X_i \cap X_j$ consists of more than one point,

(4) $x \in X_i \cap X_j \cap X_k$, where i, j, k are distinct.

<u>Definition 4</u>. Given a resolution $\tilde{V}$ of the isolated singularity $v \in V$, $\pi: \tilde{V} \to V$ satisfying the conditions of (*) we

associate a graph Γ to π as follows: To each X_i in $\pi^{-1}(v)$ assign a vertex (A_i, g_i, m_i) where g_i is the genus of X_i and m_i its self-intersection number. We join A_i to A_j by an edge if X_i meets X_j. Let S_ϵ be a small sphere around v and $K = V \cap S_\epsilon$. Clearly $\pi^{-1}(K)$ is homeomorphic to K and it is the boundary of a tubular neighborhood of $\pi^{-1}(v)$. Hence K is a singular S^1 fibration over $\pi^{-1}(v)$. In fact it is obtained by plumbing according to the graph Γ.

One can ask if there is a best resolution.

Definition 5. A resolution $\pi : \widetilde{V} \to V$ of an isolated singularity $v \in V$ is called minimal if for any resolution $\pi' : V' \to V$ there is a unique map $\sigma : V' \to \widetilde{V}$ with $\pi \circ \sigma = \pi'$. Of course the minimal resolution is unique. Brieskorn [1] proved that the minimal resolution exists if V is a surface.

Remark 6. There is a simple criterion for a resolution of a surface to be minimal. Suppose V_0 is a non-singular surface and $X \subset V_0$ is a compact irreducible curve. Then there is a non-singular surface V_1 and a proper morphism $\pi : V_0 \to V_1$ so that $\pi(X) = v \in V_1$ and π induces an isomorphism between $V_0 - X$ and $V_1 - \{v\}$ if and only if X is analytically isomorphic to $\mathbb{CP}^1$ and $(X \cdot X) = -1$. This is known as Castelnuovo's criterion. A curve X satisfying the above is called exceptional of the first kind. A resolution $\pi : \widetilde{V} \to V$ of an isolated singularity $v \in V$ is minimal if and only if no component of $\pi^{-1}(v)$ is exceptional of the first kind. Note that in general if π is the minimal resolution, then it will not necessarily satisfy the conditions of $(*)$.

Suppose $\pi : \tilde{V} \to V$ is a resolution of a normal singularity $v \in V$ and $\pi^{-1}(v) = X_1 \cup \ldots \cup X_r$, where the X_i are irreducible curves. Then the matrix $A = ((X_i \cdot X_j))$ is an important invariant of π. One can see without difficulty, Mumford [1], that A is negative definite, the diagonal entries are negative and the off diagonals are ≥ 0. It is remarkable that the converse of this theorem is true.

Theorem (Grauert). Suppose V_o is a non-singular analytic surface, $X = X_1 \cup \ldots \cup X_r$, where X_i are compact irreducible curves and $((X_i \cdot X_j))$ is negative definite. Then there is an analytic surface V_1 and a morphism $\pi : V_o \to V_1$ so that $\pi(X) = v \in V_1$ and π induces an isomorphism between $V_o - X$ and $V_1 - \{v\}$.

It is interesting to note that if V_o is algebraic V_1 need not be algebraic.

3.4. Resolution and $\mathbb{C}^*$-action

In this section we show that if V is a surface with a $\mathbb{C}^*$-action, then there is an equivariant resolution $\pi : \tilde{V} \to V$ i.e. we can choose $(\pi, \tilde{V})$ so that the $\mathbb{C}^*$ action on V extends to $\tilde{V}$.

Definition 1. Suppose G is a complex Lie group and V is an analytic space. An action σ of G on V is a morphism of analytic spaces

$$\sigma : G \times V \to V$$

so that $\sigma(gg', v) = \sigma(g, \sigma(g', v))$ and $\sigma(1, v) = v$.
We shall denote $\sigma(g, v)$ by gv when there is no danger of confusion. Recall that the action is said to be effective if $gv = v$ for all v implies $g = 1$.

Proposition 2. *Suppose* σ *is an action of* G *on* V, $I \subset \mathcal{O}_V$ *is an ideal sheaf and* $\pi : V' \to V$ *is the monoidal transform with center* I. *If* $\sigma(g)^*(I) = I$ *for all* $g \in G$ *then there is a unique action of* G *on* V' *compatible with the action on* V. *In particular if* $X \subset V$ *is invariant under the action of* G *and* π *is the monoidal transform with center* X *then the above conclusion holds.*

Proof. If $g \in G$ then g defines an automorphism $\sigma(g)$ of V. The universal property of monoidal transform (3.3) implies that if I is invariant under g there is a unique map $\tau(g)$: $V' \to V'$ so that $\pi \circ \tau(g) = \sigma(g) \circ \pi$. By the uniqueness we see that τ defines an action. To be more precise we must check that the map $\tau : G \times V' \to V'$ is analytic. Consider the diagram

$$\begin{array}{ccc} G \times V' & \overset{\tau}{\dashrightarrow} & V' \\ {\scriptstyle \pi_0}\downarrow & & \downarrow{\scriptstyle \pi} \\ G \times V & \underset{\sigma}{\longrightarrow} & V \end{array}$$

where $\pi_0 = id_G \times \pi$. Let $p_2 : G \times V \to V$ be the projection of $G \times V$ on V. Then $\sigma(g)(I) = I$ for all $g \in V$ implies $\sigma^*(I) = p_2^*(I)$. Now one can easily check that π_0 is the monoidal transform with center $p_2^*(I)$. Thus $(\sigma \circ \pi_0)^*(I)$ is locally principal and there is a unique map $\tau : G \times V' \to V'$ making the diagram commutative. This is the same as our τ above.

Proposition 3. *Suppose* σ *is an action of* G *on* V. *Then there is a unique extension of* σ *to the normalization* $\tilde{V}$ *of* V.

Proof. Just use the universal property of normalization.

Proposition 4. *Suppose G is a connected algebraic group and σ is an action of G on a surface V. Then σ leaves the following invariant:*

(1) *an isolated singular point,*

(2) *an exceptional curve,*

(3) *a singular point of an exceptional curve,*

(4) *a point $x \in V$ where two or more components of the exceptional locus meet.*

Proof. Every element $t \in G$ acts as an automorphism of V. Hence if v satisfies any of the above properties, then so does tv. But if $tv \neq v$ then the set of points satisfying that property is positive dimensional and this is impossible. If $X \subset V$ is an exceptional curve and $t(X) \neq X$, then V is covered by exceptional curves. But there are only a finite number of such curves.

3.5. Weighted Homogeneous Polynomials and Good $\mathbb{C}^*$-action

Definition 1. Suppose $(w_0,\ldots,w_n)$ are non-zero rational numbers. A polynomial $h(Z_0,\ldots,Z_n)$ is weighted homogeneous of type $(w_0,\ldots,w_n)$ if it can be expressed as a linear combination of monomials $Z_0^{i_0} \ldots Z_n^{i_n}$ for which

$$\frac{i_0}{w_0} + \ldots + \frac{i_n}{w_n} = 1$$

This is equivalent to requiring that there exist non-zero integers $q_0,\ldots,q_n$ and a positive integer d so that $h(t^{q_0}Z_0,\ldots,t^{q_n}Z_n) = t^d h(Z_0,\ldots,Z_n)$. In fact if h is weighted homogeneous of type

$(w_0,\ldots,w_n)$ then let $\langle w_o,\ldots,w_n\rangle$ denote the smallest positive integer d so that for each i there exists an integer q_i with $q_i w_i = d$. These are the q_i and d above.

Let V be the variety defined by weighted homogeneous polynomials $h_1,\ldots,h_r$ with exponents $(q_0,\ldots,q_n)$. Then there is a natural $\mathbb{C}^*$ action

$$\sigma(t,(z_0,\ldots,z_n)) = (t^{q_0}z_0,\ldots,t^{q_n}z_n) .$$

We call this action <u>good</u> if it is effective and $q_i > 0$ for all i.

<u>Proposition 2.</u> <u>Suppose</u> $V \subset \mathbb{C}^{n+1}$ <u>is an irreducible analytic variety and</u> σ <u>is a good</u> $\mathbb{C}^*$ <u>action leaving</u> V <u>invariant,</u>

$$\sigma(t,(z_0,\ldots,z_n)) = (t^{q_0}z_0,\ldots,t^{q_n}z_n) .$$

<u>Then</u> V <u>is algebraic and the ideal of polynomials in</u> $\mathbb{C}[Z_0,\ldots,Z_n]$ <u>vanishing on</u> V <u>is generated by weighted homogeneous polynomials.</u>

<u>Proof.</u> Let f belong to $\mathbb{C}\{Z_0,\ldots,Z_n\}$ the ring of convergent power series. We let f_i denote the unique <u>polynomials</u> so that

$$f(t^{q_0}Z_0,\ldots,t^{q_n}Z_n) = \sum_{i=0}^{\infty} t^i f_i(Z_0,\ldots,Z_n).$$

The power series on the right converges for sufficiently small t. Now suppose f vanishes on V near $\underline{0}$. Then $v \in V$ implies $\sum_{i=0}^{\infty} t^i f_i(v) = 0$ for all sufficiently small t. Hence $f_i(v) = 0$ for all i and all $v \in V$ near $\underline{0}$ Let $f^{(1)},\ldots,f^{(r)}$ generate the ideal $I(V)$ of all functions in $\mathbb{C}\{Z_0,\ldots,Z_n\}$ vanishing on V. Let J be the ideal generated by $\{(f^{(j)})_i\}$. Clearly $J \subset I(V)$. Now if $v \notin V$ is within the radius of convergence of $f^{(j)}$ for all j then there is some $f_i^{(j)}$ so that $f_i^{(j)}(v) \neq 0$. Hence the locus of zeros of J is V and hence the radical of J is $I(V)$. Let

J' be the ideal generated by $\{(f^{(j)})_i\}$ in $\mathbb{C}[Z_0,\dots,Z_n]$ and let I' be the radical of J'. Then $I'\mathbb{C}\{Z_0,\dots,Z_n\} = \text{rad}\, J = I(V)$. Therefore $I(V)$ is generated by polynomials.

Now let $I'(V)$ be the ideal of V in $\mathbb{C}[Z_0,\dots,Z_n]$. If $f \in I'(V)$ then $f_i \in I'(V)$. If f is a polynomial, then there are only a finite number of integers i with $f_i \neq 0$. Therefore if $f^{(1)},\dots,f^{(r)}$ generate $I'(V)$, then the weighted homogeneous polynomials $\{f_i^{(j)}\}$ generate $I'(V)$.

Proposition 3. *If $V \subset \mathbb{C}^m$ is an algebraic variety and there is a $\mathbb{C}^*$ action σ on V defined by a morphism $\sigma : \mathbb{C}^* \times V \to V$ of algebraic varieties then*

(i) *there is an embedding $j : V \to \mathbb{C}^{n+1}$ for some n and a $\mathbb{C}^*$ action $\tilde{\sigma}$ on $\mathbb{C}^{n+1}$ so that $j(V)$ is invariant and $\tilde{\sigma}$ induces σ on V,*

(ii) *by a suitable choice of coordinates in $\mathbb{C}^{n+1}$ we may write*

$$\tilde{\sigma}(t,z_0,\dots,z_n) = (t^{q_0}z_0,\dots,t^{q_n}z_n) \quad \text{where} \quad q_i \in Z ,$$

(iii) *if the action is fixed point free on $V - \{0\}$ then we may choose $q_i > 0$ for all i.*

Proof. (i) is a special case of Rosenlicht [1,Lemma 2], (ii) is proved in Chevalley [1, exposé 4, séminaire 1] and (iii) follows from Prill [1].

3.6. The Cone Over a Weighted Homogeneous Variety

Henceforth we shall assume that $V \subset \mathbb{C}^{n+1}$ and σ is a good $\mathbb{C}^*$ action leaving V invariant.

Definition 1. Let $\varphi : \mathbb{C}^{n+1} \to \mathbb{C}^{n+1}$ be defined by $\varphi(z_0,\ldots,z_n) = (z_0^{q_0},\ldots,z_n^{q_n})$ and let $V' = \varphi^{-1}(V)$. Then V' has a natural $\mathbb{C}^*$ action defined by

$$\tau(t,(z_0,\ldots,z_n)) = (tz_0,\ldots,tz_n)$$

and φ commutes with the $\mathbb{C}^*$ action. We call (φ,V') the cone over V. Note that V is the quotient of V' by $\mathbb{Z}_{q_0}\times\ldots\times\mathbb{Z}_{q_n}$ acting on $\mathbb{C}^{n+1}$ coordinatewise.

Proposition 2. The cone is a generically non-singular variety, i.e. there is an open algebraic (hence dense) subset $U_0 \subset V'$ so that if

$$I = (f_i(Z_0,\ldots,Z_n)) \qquad i = 1,\ldots,r$$

is the ideal of polynomials vanishing on V and

$$g_i(Z_0,\ldots,Z_n) = f_i(Z_0^{q_0},\ldots,Z_n^{q_n}) \qquad i = 1,\ldots,r$$

then

$$\operatorname{rank}\left(\frac{\partial g_i}{\partial z_j}\right)_w = n - s+1$$

for all $w \in U_0$ where $s = \dim_{\mathbb{C}} V$.

Proof. We may assume that V is not contained in any coordinate hyperplane $\{Z_i = 0\}$. Now V is a variety, hence it is generically non-singular i.e.

$$\operatorname{rank}\left(\frac{\partial f_i}{\partial z_j}\right)_v = n - s + 1 \qquad \text{for } v \in U\text{, open dense in } V.$$

Then

$$\left(\frac{\partial g_i}{\partial z_k}\right)_{(z_0,\dots,z_n)} = \left(\frac{\partial f_i}{\partial z_j}\right)_{(z_0^{q_0},\dots,z_n^{q_n})} \cdot \left(\frac{\partial (z_j^{q_j})}{\partial z_k}\right) .$$

There exists a point $(z_0,\dots,z_n) \in V$ with $z_i \neq 0$ for all i, so that the matrix on the right is invertible at this point. Hence

$$\operatorname{rank} \left(\frac{\partial g_i}{\partial z_k}\right)_{(z_0,\dots,z_n)} = n - s + 1 .$$

But this property holds on some open algebraic subset and the subset is non-empty. This proves the assertion.

3.7. The Quotient of $V - \{\underline{0}\}$ by $\mathbb{C}^*$

The cone V' above V is defined by homogeneous polynomials $g_1,\dots,g_r$. These polynomials define a projective variety $X' \subset \mathbb{CP}^n$. In fact X' is precisely the algebraic quotient of $V' - \{\underline{0}\}$ by $\mathbb{C}^*$. The analogue is true for V, Mumford [2, chapter 2].

Proposition 1. There is a projective variety X and an algebraic morphism $\pi : V - \{0\} \to X$ so that

(1) the fibers of π are precisely the orbits of the action,

(2) the topology of X is the quotient topology,

(3) for any open algebraic subset $U \subset X$ the algebraic functions on U are precisely the invariant functions on $\pi^{-1}(U)$.

The map $\pi' : V' - \{\underline{0}\} \to X'$ has fibers $\mathbb{C}^*$. We would like to add a zero section to get a map with fiber $\mathbb{C}$. Let

$$\Gamma_{\pi'} \subset (V' - \{\underline{0}\}) \times X'$$

be the graph of π'; let F' be the closure of Γ in $V' \times X'$ and let $\tau' : F' \to X'$ be the map induced by projection on the second factor. We have obtained F' from V' by blowing up the origin $\gamma' : F' \to V'$. Clearly $\mu'(x') = (0,x')$ gives the zero section of (τ',F'). This pair is just the hyperplane bundle of X'. Now the action of $G = \mathbb{Z}_{q_0} \times \dots \times \mathbb{Z}_{q_n}$ on V' induces an action on F'. Let F be the quotient of F' by this action. Note that F is just the closure of Γ_π in $(V-\underline{0})\times X$. The actions of $\mathbb{C}^*$ and G on V' commute, hence X is the quotient of X' by G. We have the commutative diagram

$$\begin{array}{ccc} F' & \xrightarrow{\;G\;} & F \\ {\scriptstyle\tau'}\big\downarrow\big\uparrow{\scriptstyle\mu'} & & {\scriptstyle\tau}\big\downarrow\big\uparrow{\scriptstyle\mu} \\ X' & \xrightarrow{\;G\;} & X \end{array}$$

where the horisontal maps are quotients by the action of G, μ' is the zero section, μ is the map induced by μ' and τ is the map induced by τ'. Let $\gamma : F \to V$ be the map induced by γ'.

3.8. The Canonical Equivariant Resolution of a Surface

Suppose $\dim_{\mathbb{C}} V = 2$ and V has an isolated singularity at $\underline{0}$. Then by Proposition (3.6.2) there is an open dense subset U_0 of V' so that every point of V' is simple. Hence there is an open dense subset $U \subset X'$ with the same property. Now (τ',F') is a line bundle, hence $\tau^{-1}(U)$ is non-singular. Clearly G is a finite map ramified along a finite number of fibers of τ'. Hence there is an open subset $U_1 \subset X$ so that $\tau^{-1}(U_1)$ is non-singular. Now $F - \mu(X)$ is non-singular, hence F has only a finite number of singular points along $\mu(X)$, all with neighborhoods of the form $\mathbb{C}^2/\mathbb{Z}_\alpha$ for some α. Let $\rho_0 : \tilde{V} \to F$ be the <u>minimal</u> resolution of these singular points. Then the $\mathbb{C}^*$ action extends to $\tilde{V}$ (since there is an equivariant resolution dominating $\tilde{V}$). The composite map $\rho : \tilde{V} \xrightarrow{\rho_o} F \xrightarrow{\gamma} V$ is a resolution of the singularity of V. We shall say that ρ is the <u>canonical equivariant resolution of</u> V. Since ρ is equivariant given a small $U(1)$-invariant disk D_ε at $\underline{0}$, the manifold $\rho^{-1}(D_\varepsilon)$ is a $U(1)$-invariant subset obtained by equivariant plumbing of D^2 bundles by the graph of $\rho^{-1}(\underline{0})$. Its boundary, K is therefore a smooth, orientable 3-manifold with S^1 action and $F \cup SE = \emptyset$.

The <u>proper transform</u> X_o of $X \subset F$ is the unique irreducible curve in $\tilde{V}$ so that $\rho_o(X_o) = X$. Note that the C^* action is trivial both on X and X_o. It is easily proved that the other curves of the resolution have no isotropy groups. It also follows directly from the fact that the singularity is isolated that X and X_o are isomorphic non-singular projective curves.

<u>Theorem 1</u>. <u>Let</u> $\rho^{-1}(\underline{0}) = X_o \cup \ldots \cup X_r$, <u>where</u> X_i <u>is an irreducible curve and</u> X_o <u>is the proper transform of</u> X. <u>Then</u>

(1) X_i <u>is non-singular for all</u> i, X_i <u>meets</u> X_j <u>at no more than one point,</u> X_i <u>crosses</u> X_j <u>normally at that point and</u> $X_i \cap X_j \cap X_k = \emptyset$ <u>for distinct</u> i,j,k,

(2) <u>the action is trivial on</u> X_o,

(3) <u>the action is non-trivial on</u> X_i, $i > 0$, <u>and</u> $g_i = 0, i>0$,

(4) Γ <u>is a weighted star with center</u> A_o,

(5) $m_i \leq -2$, <u>for all</u> $i > 0$.

<u>Proof</u>: By (3.4.4) we can perform a sequence of monoidal transforms with centers at fixed points of the action so that the composite $\rho' : V' \to \tilde{V}$ satisfies

(a) the action extends to V'

(b) V' and $\rho\circ\rho'$ satisfy (1).

Let $(\rho\circ\rho')^{-1}(\underline{0}) = X'_0 \cup \ldots \cup X'_r$, and let Γ' be the graph associated to $\rho\circ\rho'$. Now Γ' satisfies (2.2.2.a) and $(X'_i \cdot X'_i) < 0$ as noted in (3.3). Finally, if X'_i and X'_j have genus zero, X'_i meets X'_j and $(X'_i \cdot X'_i) = (X'_j \cdot X'_j) = -1$ then the intersection matrix $((X'_i \cdot X'_j))$ cannot be negative definite. Applying (2.2.2) we see that $g'_i = 0$ for $i > 0$ and Γ' is a weighted star with center A'_0. Thus Γ' satisfies (1) - (4). Let s be the number of $m_i = -1$. We will prove by descending induction on s that (1) - (4) are satisfied for any resolution between V' and $\tilde{V}$. Suppose X'_i is a rational curve with non-trivial action and $(X'_i \cdot X'_i) = -1$. Then by Castelnuovo's criterion (3.3.6) there is a manifold V'' and a map $f: V' \to V''$ so that $f(X'_i)$ is a point and f is an isomorphism outside of X'_i. Now X'_i meets at most two other curves, say X'_1 and X'_2. It meets each at one point and with normal crossings there. Let $\bar{X}_j = f(X'_j)$. Then $\bar{X}_1 \cdot \bar{X}_2 = f^*(\bar{X}_1) \cdot f^*(\bar{X}_2) = (X'_1 + X'_i) \cdot (X'_2 + X'_i) = 1$. Thus $\bar{X}_1$ meets $\bar{X}_2$ normally at one point. Thus V'' satisfies (1) - (4). Proceeding inductively we see that $\tilde{V}$ satisfies (1) - (4). But $\tilde{V}$ is a minimal resolution of F, hence $(X_i \cdot X_i) \leq -2$. This completes the proof.

Combining the above theorem with the results of (2.2) we obtain the main resolution theorem.

Theorem 2. The weighted graph associated to the canonical equivariant resolution of the isolated singularity of V at the origin is the star of K, $S(K)$.

Thus in order to obtain this resolution it is sufficient to find the Seifert invariants of K from the algebraic description of V.

3.9. The Seifert Invariants

Assume now that V is an algebraic surface with an isolated singularity given as the locus of zeros of some polynomials in $\mathbb{C}^{n+1}$ and it is invariant under a good $\mathbb{C}^*$ action. We shall describe how to find the Seifert invariants of K. More specific results for hypersurfaces in $\mathbb{C}^3$ are given in the next section.

1. Finding α_j. If all coordinates of a point $\underline{z} = (z_0,\ldots,z_n)$ are different from zero, then $\underline{z}$ is on a principal orbit since $(q_0,\ldots,q_n) = 1$. The point $\underline{z}$ in the hyperplane $H = \{z_{i_1} = \ldots = z_{i_k} = 0\}$ with all other coordinates non-zero has isotropy group of order $\alpha = (q_0,\ldots,\hat{q}_{i_1},\ldots,\hat{q}_{i_k},\ldots,q_n)$. The number of orbits with isotropy group $\mathbb{Z}_\alpha$ lying in H equals the number of those components of $V \cap H$ that are not in any smaller coordinate hyperplane.

2. Finding β_j. Let S be an orbit of K with isotropy group $\mathbb{Z}_\alpha$, $\alpha > 1$. For an analytic slice D^2 in K through $x \in S$ we can find an analytic isomorphism $\varphi : \Delta = \{u \in \mathbb{C} \mid |u| < 1\} \to D$ so that the induced $\mathbb{Z}_\alpha$ action τ on Δ is a standard linear action. For $\rho = \exp(2\pi i/\alpha)$ and for some $0 \le \nu < \alpha$ we have $\tau(\rho,u) = \rho^\nu u$. Then $\beta\nu \equiv 1 \bmod \alpha$ and $0 \le \beta < \alpha$. (Notice that the orientation adopted in Orlik-Wagreich [1,2] is the opposite of this.)

3. Finding b. Suppose V is invariant under the good $\mathbb{C}^*$ action

$$\sigma(t,z_0,\ldots,z_n) = (t^{q_0}z_0,\ldots,t^{q_n}z_n)$$

and d is the degree of the cone over V as defined in (3.6). Making adjustments for the present orientation convention we obtain the following formula

$$b = \frac{d}{q_0 q_1 \cdots q_n} - \sum_{j=1}^{r} \frac{\beta_j}{\alpha_j} .$$

Rather than repeating the proof as given in Orlik-Wagreich [1] we shall only outline the argument. If V is defined by homogeneous polynomials of degree d, then $q_0 = \ldots = q_n = 1$ and there are no E-orbits. In this case $V - \{0\}$ is a $\mathbb{C}^*$-bundle over X induced by the $\mathbb{C}^*$ bundle $\mathbb{C}^{n+1} - \{0\} \to CP^n$. The latter has chern class -1. The fact that X has degree d means that the map

$$H^2(CP^n; \mathbb{Z}) \to H^2(X; \mathbb{Z})$$

induced by inclusion is multiplication by d so the chern class of the bundle over X is $-d$ and therefore $b = d$ satisfying the formula in this case. The general formula is obtained as follows. Let $\varphi : V' \to V$ be the covering of V by its cone, $V = V'/G$, $G = \mathbb{Z}_{q_0} \oplus \ldots \oplus \mathbb{Z}_{q_n}$ and F, X, F', X' as in (3.7). Since V' may have non-isolated singularities the curve X' may be singular. Let $H : Y' \to X'$ be its desingularization and $F_0 = F' \times_{X'} Y'$. Since F' is a $\mathbb{C}$-bundle over X' of degree $-d$ the same holds for F_0 over Y' and $(Y' \cdot Y')_{F_0} = -d$. Let $\widetilde{V}$ be the canonical equivariant resolution and $\widetilde{X}$ the center curve. We want to compute $(\widetilde{X} \cdot \widetilde{X})_{\widetilde{V}}$. First one constructs non-singular varieties W_0 and V_1 and birational maps $\tau : W_0 \to F_0$ and $\rho_1 : V_1 \to \widetilde{V}$ and a map $\eta : W_0 \to V_1$ so that the diagram below is commutative

$$\begin{array}{ccccc} W_0 & & \xrightarrow{\quad\tau\quad} & & F_0 \\ \eta \downarrow & & & & \downarrow \pi \\ V_1 & \xrightarrow{\rho_1} & \widetilde{V} & \xrightarrow{\widetilde{\rho}} & F \end{array} .$$

Here W_o is the blowing up of the fixed points of the action of G on $Y' \subset F_o$. Then G acts freely on W_o and η is the quotient map.

Let $Y_o = \tau^{\#}(Y')$, $\tilde{X} = \tilde{\rho}^{\#}(X)$, $X_1 = \rho_1^{\#}(\tilde{X})$. The degree of the map η is $q_o q_1 \ldots q_n$ and it is easily seen that

$$(q_o \ldots q_n)(X_1 \cdot X_1)_{V_1} = (\eta^* X_1 \cdot \eta^* X_1)_{W_o} = (Y_o \cdot Y_o)_{W_o} .$$

The second part of the argument shows how the maps ρ_1 and τ change these intersection numbers. Specifically one proves that

$$(X_1 \cdot X_1)_{V_1} = (\tilde{X} \cdot \tilde{X})_{\tilde{V}}$$

and

$$(Y_o \cdot Y_o)_{W_o} + q_o \ldots q_n \sum_{j=1}^{r} \frac{\alpha_j - \beta_j}{\alpha_j} = (Y' \cdot Y')_{F_o}$$

giving the formula as asserted.

4. <u>Finding</u> g. This computation is purely algebraic. The non-singular curve X has arithmetic (and topological) genus $p_a(X) = \dim H^1(X, \mathcal{O}_X)$ which is the constant term of the Hilbert polynomial of the homogeneous coordinate ring, R_X. Now X' is defined by homogeneous polynomials so its coordimate ring, $R_{X'}$ is known. One proves that $R_X = (R_{X'}^G)^{(m)}$ where $m = q_o \ldots q_n$ and $(\ )^G$ denotes the subring fixed by G. There are technical difficulties because the ring $R_{X'}^G$ is not generated by forms of degree 1 and therefore the Hilbert polynomial is not defined, see Orlik- Wagreich [2]. An alternate method is given in (3.11) for hypersurfaces in $\mathbb{C}^3$.

3.10. Surfaces in $\mathbb{C}^3$

Suppose that V is a surface in $\mathbb{C}^3$ having an isolated singularity and admitting a good $\mathbb{C}^*$ action. It follows from (3.5.2) that V is defined by a weighted homogeneous polynomial, $h(Z_0,Z_1,Z_2)$. Using the $\mathbb{C}^*$ action it is shown in Orlik-Wagreich [1] that there is an equivariant analytic deformation of V into a surface defined by one of the following six classes of polynomials

(I) $Z_0^{a_0} + Z_1^{a_1} + Z_2^{a_2}$

(II) $Z_0^{a_0} + Z_1^{a_1} + Z_1 Z_2^{a_2}$

(III) $Z_0^{a_0} + Z_1^{a_1} Z_2 + Z_2^{a_2} Z_1$

(IV) $Z_0^{a_0} + Z_0 Z_1^{a_1} + Z_1 Z_2^{a_2}$

(V) $Z_0^{a_0} Z_1 + Z_1^{a_1} Z_2 + Z_0 Z_2^{a_2}$

(VI) $Z_0^{a_0} + Z_1 Z_2$

inducing an equivariant diffeomorphism of respective neighborhood bounderies of the isolated singularity at the origin.

Thus it is sufficient to study these six classes of polynomials. The polynomial $Z_0^{a_0} + Z_1 Z_2$ is analytically isomorphic to $Z_0^{a_0} + Z_1^2 + Z_2^2$ so it may be treated as a subclass of I .

Assuming that the weights equal w_i , $i = 0,1,2$ and they are reduced as a fraction to $w_i = u_i/v_i$, we introduce auxiliary integers

$$c = (u_0,u_1,u_2)$$

$c_0 = (u_1,u_2)/c$, $c_1 = (u_0,u_2)/c$, $c_2 = (u_0,u_1)/c$, $c_{1,2} = u_0/cc_1c_2$, $c_{0,2} = u_1/cc_0c_2$, $c_{0,1} = u_2/cc_0c_1$. Note that c_0, c_1,c_2 are pairwise relatively prime, $c_{0,1},c_{0,2}$ and $c_{1,2}$ are pairwise relatively prime and $(c_i,c_{j,k}) = 1$ if i,j and k are distinct.

The integer d defined as the least common multiple of the u_i equals

$$d = cc_o c_1 c_2 c_{o,1} c_{o,2} c_{1,2}$$

and from this we compute $q_i = d/w_i$ as $q_0 = v_0 c_0 c_{0,1} c_{0,2}$, $q_1 = v_1 c_1 c_{o,1} c_{1,2}$, $q_2 = v_2 c_2 c_{0,2} c_{1,2}$.

1. Orbits with non-trivial isotropy groups are in the hyperplane sections. The number of orbits in a given hyperplane section is the number of irreducible components of the curve of intersection. For example in class I the subset

$$\{z_0 = 0,\ z_1^{a_1} + z_2^{a_2} = 0\} \cap S^5$$

has isotropy group $\mathbb{Z}_{\alpha_0} = \mathbb{Z}_{(q_1,q_2)} = \mathbb{Z}_{c_{1,2}}$. It consists of $n_0 = (a_1,a_2) = cc_0$ orbits. Similar arguments yield the following table where α_0, α_1, α_2 are the three possible isotropy groups in the three hyperplane sections and n_0, n_1, n_2 are the number of orbits in each.

	α_0	n_0	α_1	n_1	α_2	n_2
I	$c_{1,2}$	cc_0	$c_{0,2}$	cc_1	$c_{0,1}$	cc_2
II	$c_{1,2}$	$(cc_0-1)/v_2$	$v_2c_{1,2}$	1	$c_{0,1}$	c
III	$c_{1,2}$	$(cc_0-v_1-v_2)/v_1v_2$	$v_2c_{1,2}$	1	$v_1c_{1,2}$	1
IV	$c_{0,1}$	$(c-1)/v_1$	v_2	1	$v_1c_{0,1}$	1
V	v_0	1	v_1	1	v_2	1

2. In order to compute β we note that a sufficiently close slice in V maps diffeomorphically onto a slice in K so we may consider the former. All orbits in the same hyperplane section have the same orbit type since so does the whole hyperplane. Consider for example an orbit with isotropy group $\mathbb{Z}_{\alpha_0}$ in class I as above. Let $\xi = \exp(2\pi i/\alpha_0)$. The action of ξ in $\mathbb{C}^3$ is

$$\xi(z_0,z_1,z_2) = (\xi^{q_0}z_0,z_1,z_2) .$$

Considering the z_0 plane as a slice the action is the standard action of type $[\alpha_0,q_0]$ and hence β_0 is defined by the congruence

$$q_0\beta_0 \equiv 1(\text{mod }\alpha_0) .$$

Notice that this is the orientation convention of (1.1.7) and the opposite of that used in Orlik-Wagreich [1,2]. For an orbit on the intersection of two hyperplanes, e.g. in class II

$$\{z_0 = z_1 = 0,\ |z_2|^2 = 1\}$$

the slice at $z_2 = 1$ is the curve $\{z_0^{a_0} + z_1^{a_1} + z_1 = 0\}$. This curve near $(0,0,1)$ may be "approximated" by changing it by an analytic automorphism

$$\varphi(z_0,z_1) = (z_0 + h_0(z_0,z_1),\ z_1 + h_1(z_0,z_1))$$

where $h_i \in \mathbb{C}\{z_0,z_1\}$ have all terms of degree ≥ 2 . The curve $\{z_0^{a_0} + z_1 = 0\}$ is an approximation and if $\xi = \exp(2\pi i/\alpha_1)$ the action in the slice is approximated by

$$\xi(z_0,-z_0^{a_0},1) = (\xi^{q_0}z_0,-\xi^{q_0a_0}z_0^{a_0},1) = (\xi^{q_0}z_0,-z_0^{a_0},1) .$$

So we have $\nu_1 = q_0$ and hence

$$\beta_1 q_0 \equiv 1(\text{mod }\alpha_1) .$$

The table below gives the ν_j , $j = 0,1,2$. Since $\beta_j\nu_j \equiv 1(\text{mod }\alpha_j)$ and $0 \leq \beta_j < \alpha_j$ this determines the β_j .

	ν_0	ν_1	ν_2
I	q_0	q_1	q_2
II	q_0	q_0	q_2
III	q_0	q_0	q_0
IV	q_2	q_0	q_2
V	q_2	q_0	q_1

3. As we have mentioned earlier b is given by the formula

$$b = \frac{d}{q_0 q_1 q_2} - \sum_{j=1}^{r} \frac{\beta_j}{\alpha_j} .$$

4. Finally the construction of the previous section gives the following expression for g, Orlik-Wagreich [1,(3.5.1);2,(5.4)]

$$2g = \frac{d^2}{q_0 q_1 q_2} - \frac{d(q_0,q_1)}{q_0 q_1} - \frac{d(q_1,q_2)}{q_1 q_2} - \frac{d(q_2,q_0)}{q_2 q_0}$$
$$+ \frac{(d,q_0)}{q_0} + \frac{(d,q_1)}{q_1} + \frac{(d,q_2)}{q_2} - 1 .$$

We shall give an alternate way of obtaining this formula using the fibration theorem of Milnor [1] in the next section. First consider an example.

Let a variety V in $\mathbb{C}^3$ be defined by the weighted homogeneous polynomial of class III, $h(Z) = Z_0^{15} + Z_1^4 Z_2 + Z_2^7 Z_1$. It has an isolated singularity at the origin. We find $w_0 = 15$, $w_1 = 9/2$, $w_2 = 9$, $d = 45$, $q_0 = 3$, $q_1 = 10$, $q_2 = 5$, $c = 3$, $c_0 = 3$, $c_{1,2} = 5$ and the other c-s equal 1. The locus $\{z_0 = 0,\ z_1^3 + z_2^6 = 0\} \cap S^5$ consists of 3 orbits with stability group of order $\alpha_0 = (q_1,q_2) = 5$. There is one orbit $\{z_0 = z_1 = 0\} \cap S^5$ with $\alpha_1 = q_2 = 5$ and one orbit $\{z_0 = z_2 = 0\} \cap S^5$

with $\alpha_2 = q_1 = 10$. The corresponding $\nu_0 = \nu_1 = \nu_2 = q_0$ so $\beta_0 = 2$, $\beta_1 = 2$ and $\beta_2 = 7$. The formula for b gives $b = -1$ and the formula for g gives $g = 3$. Thus

$$K = \{-1;(0,3,0,0);(5,2),(5,2),(5,2),(5,2),(10,7)\}$$

and the star of K

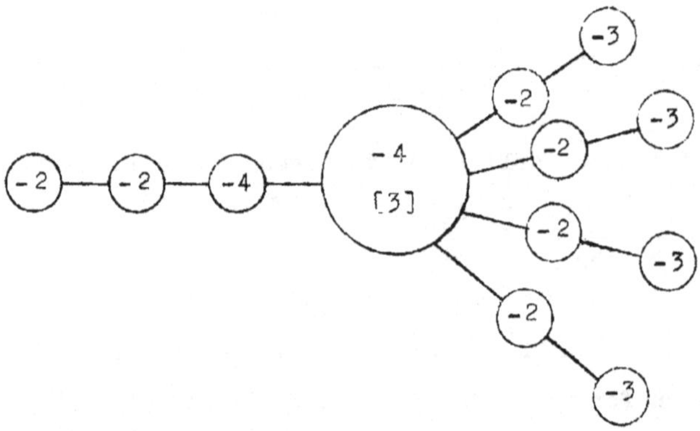

is the dual of the graph of the canonical equivariant resolution of the singularity of V .

3.11. Milnor's Fibration Theorem

Let V be an algebraic hypersurface in $\mathbb{C}^{n+1}$ defined by the zeros of a polynomial, $V = \{\underline{z} \mid f(\underline{z}) = 0\}$. Let $\underline{x}$ be an arbitrary point on V and S_ε a sufficiently small sphere centered at $\underline{x}$. Let $K = V \cap S_\varepsilon$. The following fibration theorem is due to Milnor [1].

__Theorem.__ __The mapping__

$$\Phi(\underline{z}) = f(\underline{z})/|f(\underline{z})|$$

__from__ $S_\varepsilon - K$ __to__ S^1 __is the projection map of a smooth fiber__ bundle. Each fiber

$$F_\theta = \Phi^{-1}(e^{i\theta}) \subset S_\epsilon - K$$

is a smooth parallelizable $2n$-manifold.

For an isolated singularity there is additional information.

Theorem. If x is an isolated critical point of f then each fiber F_θ has the homotopy type of a bouquet $S^n v \dots v S^n$ of n-spheres. Their number, μ is strictly positive. Each fiber can be considered as the interior of a smooth compact manifold with boundary

$$\text{closure}(F_\theta) = F_\theta \cup K$$

where the common boundary K is an $(n-2)$-connected smooth $(2n-1)$-manifold.

The complement of K in S_ϵ, $S_\epsilon - K$ is therefore obtained from $F \times [0,2\pi]$ by identifying F_0 and $F_{2\pi}$ by a homeomorphism

$$h : F \to F ,$$

called the characteristic map. The Wang sequence associated to this fibration is according to Milnor [1,8.4]

$$\dots \to H_{j+1}(S_\epsilon - K) \to H_jF \xrightarrow{h_*-I_*} H_jF \to H_j(S_\epsilon - K) \to \dots$$

where I is the identity map of F. In case x is an isolated singularity we can use the information on the connectivity of F and K, Alexander duality and Poincaré duality to see that for $n \geq 2$ the Wang sequence reduces to the short exact sequence

$$0 \to H_nK \to H_nF \xrightarrow{h_*-I_*} H_nF \to H_{n-1}K \to 0 .$$

Let $\Delta(t) = \det(tI_*-h_*)$ denote the characteristic polynomial of the transformation $h_* : H_nF \to H_nF$.

If $f(\underline{z})$ is a weighted homogeneous polynomial of type $(w_0,\ldots,w_n)$ then Milnor shows furthermore that F is diffeomorphic to the non-singular algebraic variety

$$F' = \{\underline{z} \mid f(\underline{z}) = 1\}$$

and the characteristic map h may be chosen

$$h(z_0,\ldots,z_n) = (\xi^{q_0} z_0,\ldots,\xi^{q_n} z_n)$$

where $\xi = \exp(2\pi i/d)$. In particular h is of finite order divisible by d . Thus the minimal polynomial of h_* divides $(t^d - 1)$ and hence it is a square-free polynomial. This implies in turn that the rank of the kernel and cokernel of (h_*-I_*) equals the exponent $\varkappa$ of $(t-1)$ in $\Delta(t)$. An expression for $\varkappa$ was obtained by Milnor-Orlik [1] in terms of the weights. Let $w_i = u_i/v_i$, $i = 0,\ldots,n$ be in irreducible form. Given integers $a_0,\ldots,a_k$ denote their least common multiple by $[a_0,\ldots,a_k]$. We have

$$\varkappa(w_0,\ldots,w_n) = \sum (-1)^{n-s} \frac{w_{i_0}\cdots w_{i_s}}{[u_{i_0},\ldots,u_{i_s}]}$$

where the sum is taken over the 2^{n+1} subsets $\{i_0,\ldots,i_s\}$ of $\{0,\ldots,n\}$.

In the case of a surface in $\mathbb{C}^3$ we already know H_1K in terms of generators and relations. There are $2g$ free generators from the partial cross section together with the generators

$$q_0, q_1,\ldots,q_r, h$$

satisfying the relations:

$$q_0 + q_1 + \ldots + q_r = 0$$

$$q_0 + bh = 0$$

$$\alpha_j q_j + \beta_j h = 0 \qquad\qquad j = 1,\ldots,r .$$

The first comes from the partial cross section and the remaining ones from the sewings of the solid torus neighborhoods of the b-obstruction and the E-orbits. The determinant of the relation matrix equals $p = b\alpha_1 \dots \alpha_r + \beta_1\alpha_2 \dots \alpha_r + \dots + \alpha_1\alpha_2 \dots \beta_r$

$$\frac{p}{\alpha_1 \dots \alpha_r} = b + \sum_{j=1}^{r} \frac{\beta_j}{\alpha_j} .$$

On the other hand from the expression for b (3.10.3) we obtain

$$b + \sum_{j=1}^{r} \frac{\beta_j}{\alpha_j} = \frac{d}{q_0 q_1 q_2}$$

so we see that $p > 0$ and therefore the generators $q_0, \dots, q_r, h$ are torsion elements of H_1K. Thus

$$\varkappa(w_0, w_1, w_2) = \text{rank } H_1K = 2g .$$

Substituting $w_i = d/q_i$, $i = 0,1,2$ in $\varkappa(w_0, w_1, w_2)$ yields (3.10.4).

Although this proof is correct it is somewhat unsatisfactory in that the essential reason for $p > 0$ is hidden in the proof of the formula for b. Examining that proof one observes that $p > 0$ is equivalent to the negative definiteness of the quadratic form of the resolution.

Finally note that this approach is valid only for hypersurfaces. For higher embedding dimensions the algebraic method mentioned in (3.9) has no topological replacement at present.

3.12. Non-isolated Singularities

Rather than giving a detailed account of the resolution of non-isolated singularities of surfaces with a good $\mathbb{C}^*$ action as in Orlik-Wagreich [2] we shall point out the additional difficulties compared with the isolated case.

1. Let $\theta : \bar{V} \to V$ be the normalization (3.3.3) of V, where $V \subset \mathbb{C}^{n+1}$ is a surface invariant under a good $\mathbb{C}^*$ action. We are interested in the resolution of the isolated singularities of $\bar{V}$ using the methods already developed. The fact that V is given with a good $\mathbb{C}^*$ action is of little help, however, because the same may not be assumed of $\bar{V}$. A canonical equivariant resolution of the singularities of $\bar{V}$ may be constructed as follows: Let V' be the cone over V in $\mathbb{C}^{n+1}$ and $V'-0/\mathbb{C}^* = X' \subset CP^n$. Let $\eta : \bar{X}' \to X'$ be the normalization (resolution) of the projective curve X'. Let F' denote the hyperplane (Hopf) bundle of CP^n restricted to X'. Since the degree of F' is negative Grauert's Theorem (3.3) assures that there is a birational map $j' : F' \to V'$ collapsing the zero section. Let $\bar{F}' = \eta^*(F')$ and $\bar{\gamma}' = \bar{F}' \to \bar{V}'$ be the map collapsing the zero section. Now $\bar{V}'$ maps into the normalization of V' and it is normal so it is the normalization. $\bar{F}'$ is non-singular and the action of $G = \mathbb{Z}_{q_0} \oplus \ldots \oplus \mathbb{Z}_{q_n}$ on F' extends. Let $\bar{F} = \bar{F}'/G$, $\bar{V} = \bar{V}'/G$ and $\bar{\gamma} = \bar{F} \to \bar{V}$ the induced map. Finally let $\tilde{\rho} : \tilde{V} \to \bar{F}$ be the minimal resolution of the quotient singularities of $\bar{F}$. Then $\rho = \bar{\gamma}\tilde{\rho} : \tilde{V} \to \bar{V}$ is the canonical equivariant resolution of $\bar{V}$.

2. Since the action extends, $\bar{V}$ has an isolated singularity at the origin whose resolution is determined by the Seifert invariants of $\bar{K}$. The topology of V at the origin is determined by the map $\theta|_{\bar{K}} : \bar{K} \to K$. In general K is not a manifold and θ may identify orbits of $\bar{K}$, some by maps of different degrees. One needs some notation for these objects and an equivariant classification theorem.

3. The central object is obtaining the Seifert invariants of $\bar{K}$ and understanding the map θ from the algebraic description of V.

The isotropy groups of orbits in K are easy to read off. The slice at $\underline{z} \in K$ may consist of several disks meeting at $\underline{z}$. The number of orbits mapping onto the orbit of $\underline{z}$ is determined by the number of orbits of the action of $\mathbb{Z}_\alpha$ in the slice. If k disks of the slice are mapped into each other by $\mathbb{Z}_\alpha$, then there is one orbit with isotropy group $\mathbb{Z}_{\alpha/k}$ in $\bar{K}$ mapping onto the orbit of $\underline{z}$ by a map of degree k. The action of $\mathbb{Z}_{\alpha/k}$ in the individual slice determines β (as an invariant of $\bar{K}$). The obstruction class b is obtained by the same formula as before. The genus $g(\bar{X})$ of the non-singular curve $\bar{X} = \bar{V} - \underline{0}/\mathbb{C}^*$ is obtained from the arithmetic genus $p_a(X)$ of the (possibly singular) curve $X = V - 0/\mathbb{C}^*$ using the formula

$$g(\bar{X}) = p_a(X) - \sum_{x \in X} \delta_x$$

where δ_x is an invariant of the singular point $x \in X$. The computations are, of course, harder. They are carried out for hypersurfaces of $\mathbb{C}^3$ in Orlik-Wagreich [2].

4. Equivariant Cobordism and the α-Invariant

This chapter is a brief extract from the thesis of Ossa [1]. First some general notation is introduced then the basic facts about S^1-manifolds are given. Next the fixed point free cobordism group of oriented, closed, smooth 3-dimensional fixed point free S^1-manifolds is discussed in detail. It is shown to be free and generators are constructed. An algorithm for finding the cobordism class in terms of these generators from the Seifert invariants is also obtained.

Using a fixed point theorem in Atiyah-Singer [1], an invariant is defined for fixed point free circle actions. It is a rational function in $Q(t)$. This invariant is computed for 3-dimensional S^1-manifolds.

4.1. Basic Results

All manifolds and bundles are assumed smooth and orientable. Given the vector bundles $\eta_1 \to X_1$, $\eta_2 \to X_2$ define $\eta_1 \hat{\oplus} \eta_2$ by the Whitney sum of the pullbacks of the projections $pr_i : X_1 \times X_2 \to X_i$, $i = 1,2$. as

$$\eta_1 \hat{\oplus} \eta_2 = pr_1^* \eta_1 \oplus pr_2^* \eta_2 .$$

Let G be a compact Lie group, H a closed subgroup and $(H) = \{gHg^{-1} \mid g \in G\}$. A family of subgroups F is called admissible if $H \in F$ implies $(H) \subset F$. All families of subgroups will be assumed admissible. Let M^n be a G-manifold and assume that G is orientation preserving. M is called of type (F,F') if $p \in M$ then $G_p \in F$ for all $p \in M$ and if $p \in \partial M$ then $G_p \in F'$ for all $p \in \partial M$. It is called (F,F')-bounding if there is an (F,F)-manifold W^{n+1} so that M is an equivariant submanifold of ∂W and for every point $p \in \partial W - M$, $G_p \in F'$. We also call W an (F,F')-cobordism for M. Two G-manifolds M_1 and M_2 of type (F,F') are (F,F')-cobordant if the disjoint union $M_1 + (-M_2)$ is (F,F')-bounding. This is an equivalence relation. Denote by $\mathcal{O}_n(G;F,F')$ the equivalence classes of n-dimensional G-manifolds of type (F,F') and $\mathcal{O}_*(G;F,F') = \bigoplus_n \mathcal{O}_n(G;F,F')$.

Let $F \supset F' \supset F''$ be families of subgroups of G. Then there is an exact sequence

$$\cdots \to \mathcal{O}_n(G;F',F'') \xrightarrow{i} \mathcal{O}_n(G;F,F'') \xrightarrow{j} \mathcal{O}_n(G;F,F') \xrightarrow{\partial} \mathcal{O}_{n-1}(G;F',F'') \to \cdots$$

where i and j are induced by inclusion and ∂ is restriction to the boundary.

A G-vector bundle of dimension (k,n) is defined as a smooth G-vector bundle with fiber dimension k over a smooth, closed n-manifold. Assume that the total space is orientable and the action of G is orientation preserving. It will be called of type (F,H) if

(i) each isotropy group of the zero section contains a subgroup conjugate to H,

(ii) each isotropy group of the associated sphere bundle is in $F - (H)$.

A G-vector bundle ξ of type (F,H) bounds if there is a G-vector bundle η with oriented total space over a manifold with boundary so that ξ is equivariantly diffeomorphic to the restriction of η to the boundary of its base. Two G-vector bundles ξ and ξ' of type (F,H) are (F,H)-cobordant if the disjoint union $\xi + (-\xi')$ bounds. Again, (F,H)-bounding is an equivalence relation and the collection of equivalence classes $\psi_n^k(G;F,H)$ forms an abelian group under disjoint union. Let $\psi_*^*(G;F,H) = \bigoplus_{k,n} \psi_n^k(G;F,H)$. Note that $\psi_*^{2k+1}(G;F,H) = 0$ follows from the orientation assumption, e.g. if G is abelian.

Given a G-manifold M'' of type $(F,F-(H))$ the set of points $p \in M$ so that G_p contains a conjugate of H is a closed G-invariant submanifold of $M - \partial M$. Let ξ be its normal bundle in M. Then ξ is a G-vector bundle of type (F,H). It is easily seen that the map $M \to \xi$ induces an Ω_* module isomorphism

$$\mathcal{O}_n(G;F,F-(H)) \longrightarrow \bigoplus_k \psi_{n-2k}^{2k}(G;F,H) .$$

The inverse map is given by taking the associated disk bundle of ξ.

4.2. Fixed Point Free S^1-Actions

Let F_m be the family of subgroups of S^1 with order $\leq m$, $F_\infty = \bigcup_m F_m$ and F_S all subgroups of S^1. Note that $\mathbb{Z}_m$ in F_m and S^1 in F_S are maximal elements. Let us use the simplified notation

$$\begin{aligned} \mathcal{O}_n(m) &= \mathcal{O}_n(S^1;F_m,\emptyset) \\ \mathcal{O}_n(\infty) &= \mathcal{O}_n(S^1;F_\infty,\emptyset) \\ \mathcal{O}_n(S^1) &= \mathcal{O}_n(S^1;F_S,\emptyset) \end{aligned}$$

and similarly

$$\begin{aligned} \psi_n^k(m) &= \psi_n^k(S^1;F_m,\mathbb{Z}_m) \\ \psi_n^k(S^1) &= \psi_n^k(S^1;F_S,S^1) . \end{aligned}$$

Let M be an S^1-manifold and $H \subset S^1$ a closed subgroup. Define $I(H) = \{p \in M \mid h(p) = p,\ \forall h \in H\}$. Clearly $I(H)$ is an invariant submanifold in M. Let $N(H)$ be its normal bundle.

We call M an S^1-manifold with <u>complex normal bundles</u> if for every H the bundle $N(H)$ has the structure of a complex S^1-vectorbundle satisfying the condition that if $H_1 \subset H_2$ then the bundle $N(H_1)|I(H_2)$ is a complex S^1-subbundle of $N(H_2)$. The corresopnding cobordism groups are denoted by $\bar{\mathcal{O}}_n(m)$, $\bar{\mathcal{O}}_n(\infty)$ and $\bar{\mathcal{O}}_n(S^1)$. Similarly we define complex vector bundles of type (m) over oriented S^1-manifolds where the operation of S^1 is compatible with the complex structure to obtain the groups $\bar{\Psi}_n^k(m)$ of complex k-dimensional vector bundles of type (m) over n-manifolds. This yields the exact sequence

$$\cdots \to \bar{\mathcal{O}}_n(m-1) \to \bar{\mathcal{O}}_n(m) \to \bigoplus_k \bar{\Psi}_{n-2k}^k(m) \to \bar{\mathcal{O}}_{n-1}(m-1) \to \cdots .$$

Given a complex representation r of $\mathbb{Z}_m$ with no trivial summand we can form the cobordism group $\bar{\Psi}_n(m,r)$ of complex S^1-vector bundles of type $(\mathbb{Z}_m, r)$ over oriented S^1-manifolds. Let $\bar{R}^k(\mathbb{Z}_m)$ denote the set of equivalence classes of complex k-dimensional representations of $\mathbb{Z}_m$ with no trivial summand. Clearly

$$\bar{\Psi}_n^k(m) = \bigoplus_{r \in \bar{R}^k(\mathbb{Z}_m)} \bar{\Psi}_n(m,r)$$

<u>Lemma 1</u>. <u>Let</u> $r: \mathbb{Z}_m \to U(k)$ <u>be a complex representation of</u> $\mathbb{Z}_m$ <u>with no trivial summand</u>. <u>Let</u> $\bar{C}(r)$ <u>be the centralizer of</u> $r(\mathbb{Z}_m)$ <u>in</u> $U(k)$. <u>Then there is a canonical</u> Ω_* <u>module isomorphism with the singular bordism group of Conner-Floyd</u> [1]

$$\bar{\Psi}_n(m,r) = \Omega_{n-1}[B(S^1/\mathbb{Z}_m) \times B(\bar{C}(r))] .$$

<u>Proof</u>. Let $\xi \in \bar{\Psi}_n(m,r)$ and let $\tilde{\xi}$ denote the associated principal $U(k)$ bundle. Now S^1 operates on the left on ξ and

$U(k)$ on the right on $\tilde{\xi}$. Let

$$\eta = \{e \in \tilde{\xi} \mid he = er(h) , \forall h \in \mathbb{Z}_m\} .$$

Then S^1 acts on η from the left. $\bar{C}(r)$ operates as a subgroup of $U(k)$ on the right on $\tilde{\xi}$ and hence on η. Define a left action of $\bar{C}(r)$ on η by $\sigma e = e\sigma^{-1}$. This gives a left action of $S^1 \times \bar{C}(r)$ on η. Define

$$\Delta = \{(h,r(h)) \mid h \in \mathbb{Z}_m\}$$

a normal subgroup of $S^1 \times \bar{C}(r)$. It is easily seen that Δ is exactly the isotropy group of every point of η under the action of $S^1 \times \bar{C}(r)$ and η is a principal $S^1 \times \bar{C}(r)/\Delta$ bundle with base M/S^1 defining an element of $\Omega_{n-1}[B(S^1 \times \bar{C}(r))/\Delta]$ and it follows that

$$S^1 \times \bar{C}(r)/\Delta \cong S^1/\mathbb{Z}_m \times \bar{C}(r) .$$

Conversely, given a principal $S^1/\mathbb{Z}_m \times \bar{C}(r)$ bundle η over M/S^1, we obtain the principal $U(k)$ bundle $\tilde{\xi}$ with S^1 action over M by noting that there is a canonical map $\nu : \eta \times U(k) \to \tilde{\xi}$ given by $(e,\sigma) \to e\sigma$ equivariant with respect to the S^1 action. It is surjective and $\nu(e_1,\sigma_1) = \nu(e_2,\sigma_2)$ iff $\sigma_1\sigma_2^{-1} \in \bar{C}(r)$ and $e_2 = e_1\sigma_1\sigma_2^{-1}$. Thus $\tilde{\xi}$ is the quotient of $\eta \times U(k)$ by the action of $\bar{C}(r)$ given by $\sigma(e,s) = (e\sigma^{-1},\sigma s)$.

Let $\xi_n \to CP^n$ be the Hopf bundle. Then the Ω_* algebra $\oplus_k \Omega_*(BU(k))$ is a polynomial algebra generated by the classes $[\xi_n]$, $n \geq 0$. According to Conner-Floyd [2,(18.1)] one has to show that if for a k-tuple $\omega = (n_1,\ldots,n_k)$, $n_1 \geq n_2 \geq \cdots \geq n_k \geq 0$ we associate the bundle $\xi_\omega = \xi_{n_1} \hat{\oplus} \cdots \hat{\oplus} \xi_{n_k}$ over $P_\omega = CP^{n_1} \times \cdots \times CP^{n_k}$ with the classifying map f_ω, then the classes $f_{\omega *}[P_\omega] \in H_*(BU(k);\mathbb{Z})$ form a $\mathbb{Z}$-basis for $H_*(BU(k);\mathbb{Z})$. This is done by the usual characteristic class argument.

Recall that every complex representation $r : \mathbb{Z}_m \to U(k)$ is a sum of linear representations. Denote by $r_j : \mathbb{Z}_m \to U(1)$, $j = 1,\ldots,m-1$ the representation that sends the generator $\exp(2\pi i/m)$ of $\mathbb{Z}_m$ to $\exp j\,(2\pi i/m)$. Let kr_j denote the k-fold direct sum of r_j. Then for some non-negative $k_1,\ldots,k_{m-1}$ with $k_1+\ldots+k_{m-1} = k$ the representation r is equivalent to $k_1r_1 \oplus \ldots \oplus k_{m-1}r_{m-1}$. Thus $\bar{C}(r)$ is isomorphic to $U(k_1)\times\ldots \times U(k_{m-1})$ and since $S^1/\mathbb{Z}_m \cong S^1$, we have from Lemma 1:

$$\bar{\Psi}_n(m,r) = \Omega_{n-1}(BS^1 \times BU(k_1) \times \ldots \times BU(k_{m-1})) .$$

Since $H_*(BU(k);\mathbb{Z})$ has no odd torsion, the Künneth formula of singular bordism theory applies, Conner-Floyd [2,(44.1)] and one obtains the following explicit generators. Let S_m^{2q-1} denote the $(2q-1)$ sphere $\{(z_1,\ldots,z_q) \in \mathbb{C}^q \mid \Sigma z_i\bar{z}_i = 1\}$ with the ineffictive S^1 action $t(z_1,\ldots,z_q) = (t^m z_1,\ldots,t^m z_q)$. Let $\xi_n^{(j)}$ denote the Hopf bundle over CP^n with S^1 acting by multiplication by t^j in each fiber.

Theorem 2. *$\bar{\Psi}_*^k(m) = \oplus_n \bar{\Psi}_n^k(m)$ is freely generated as an Ω_* module by*

$$S_m^{2q-1} \times (\xi_{n_1}^{(j_1)} \hat{\oplus} \ldots \hat{\oplus} \xi_{n_k}^{(j_k)})$$

where $q \geq 1$; $m-1 \geq j_1 \geq j_2 \geq \ldots \geq j_k \geq 1$ *and* $n_s \geq n_{s+1}$ *if* $j_s = j_{s+1}$.

Theorem 3. (a) *The canonical Ω_* module homomorphism*

$$i : \bar{\mathcal{O}}_*(m-1) \to \bar{\mathcal{O}}_*(m)$$

is injective.

(b) $j : \bar{\mathcal{O}}_*(m) \to \oplus \bar{\Psi}_*^k(m)$ *is surjective.*

(c) $\bar{\mathcal{O}}_*(m)$ *is freely generated as an* Ω_*

module by

$$S(\xi_{n_o}^{(j_o)} \hat{\oplus} \xi_{n_1}^{(j_1)} \hat{\oplus} \dots \hat{\oplus} \xi_{n_s}^{(j_s)})$$

where $s \geq 0$, $m \geq j_o > j_1 \geq \dots \geq j_s \geq 1$ and $n_\sigma \geq n_{\sigma+1}$ if $j_\sigma = j_{\sigma+1}$.

Here $S(\eta)$ denotes the sphere bundle of the bundle η.

Proof. If η_1 and η_2 are of type (S^1) so that every isotropy group in $S(\eta_1)$ is $\mathbb{Z}_m$ and in $S(\eta_2)$ of order $< m$, then $S(\eta_1 \hat{\oplus} \eta_2)$ is of type (m) and the normal bundle $N(\mathbb{Z}_m)$ of the fixed set $I(\mathbb{Z}_m)$ is equivariantly equivalent to $S(\eta_1) \times \eta_2$. In the exact sequence

$$\dots \to \bar{\mathcal{O}}_n(m-1) \xrightarrow{i} \bar{\mathcal{O}}_n(m) \xrightarrow{j} \underset{k}{\oplus} \bar{\Psi}^k_n(m) \xrightarrow{\partial} \bar{\mathcal{O}}_{n-1}(m-1) \to \dots$$

$\underset{k}{\oplus} \bar{\Psi}^k_*(m)$ is free on the generators given in Theorem 2. The element of $\bar{\mathcal{O}}_*(m)$

$$S(\xi_{q-1}^{(m)} \hat{\oplus} \xi_{n_1}^{(j_1)} \hat{\oplus} \dots \hat{\oplus} \xi_{n_k}^{(j_k)})$$

maps onto the corresponding generator by the remark above so j is surjective and by exactness i is injective. Part (c) follows from induction on m.

In particular one obtains the following.

Corollary 4. $\bar{\mathcal{O}}_*(\infty)$ is freely generated as an Ω_* module by

$$S(\xi_{n_o}^{(j_o)} \hat{\oplus} \xi_{n_1}^{(j_1)} \hat{\oplus} \dots \hat{\oplus} \xi_{n_s}^{(j_s)})$$

where $s \geq 0$, $j_o > j_1 \geq j_2 \geq \dots \geq j_s \geq 1$ and $n_\sigma \geq n_{\sigma+1}$ if $j_\sigma = j_{\sigma+1}$.

4.3. 3-Manifolds

The cobordism group of 3-dimensional fixed point free S^1-manifolds is determined as follows.

Theorem 1. $\mathcal{O}_3(\infty)$ is free abelian with free generators

$$S(\xi_0^{(j_0)} \hat{\oplus} \xi_0^{(j_1)}) , \quad j_0 \geq 2j_1 .$$

Proof. Consider the relations:

(i) $[S(\xi_0^{(m)} \hat{\oplus} \xi_0^{(n)})] = [S(\xi_0^{(m+n)} \hat{\oplus} \xi_0^{(m)})] + [S(\xi_0^{(m+n)} \hat{\oplus} \xi_0^{(n)})]$, $m,n \geq 1$

(ii) $[S(\xi_1^{(j)})] = 2[S(\xi_0^{(2j)} \hat{\oplus} \xi_0^{(j)})]$, $j \geq 1$.

The first is obtained from the S^1 action on CP^2 given by $t[z_0:z_1:z_2] = [z_0:t^m z_1:t^{m+n}z_2]$ observing that the fixed point set consists of the three points $[1:0:0]$, $[0:1:0]$ and $[0:0:1]$ and the above are their normal sphere bundles. The second follows by noting that $S(\xi_1^{(j)}) = S(\xi_0^{(j)} \hat{\oplus} \xi_0^{(j)})$ and letting $m = n = j$ in (i). Thus it follows from (4.2.4) that the image of

$$\varphi\colon \bar{\mathcal{O}}_3(\infty) \to \mathcal{O}_3(\infty)$$

is generated by the above generators. In order to prove that φ is an isomorphism we first claim that φ is onto. This means that every 3-dimensional orientable fixed point free S^1-manifold has complex normal bundles. This is obvious since these are oriented D^2-bundles over S^1 . To show that φ is injective it is enough to show that the generators given in the theorem are linearly independent in $\mathcal{O}_3(\infty)$. Here is an outline of this argument. Using (ii) it suffices to prove that if Y is an oriented 4-dimensional fixed point free S^1-manifold with boundary

$$\partial Y = \sum_{\substack{j_0 > 2j_1 \\ j \geq 1}} a_{j_0, j_1} S(\xi_0^{(j_0)} \hat{\oplus} \xi_0^{(j_1)}) + \sum_{j \geq 1} b_j S(\xi_1^{(j\,)})$$

then the coefficients a_{j_0, j_1} and b_j are zero. First it is shown that Y is cobordant to Y' where Y' is a fixed point free S^1- manifold with complex normal bundles and $\partial Y = \partial Y'$. Using (4.2.3a) and a downward induction on the orders of the isotropy groups one obtains the announced result.

Next we shall express the cobordism class of an arbitrary oriented fixed point free S^1-manifold

$$M = \{b; (o,g,0,0); (\alpha_1, \beta_1), \ldots, (\alpha_r, \beta_r)\}$$

in terms of the generators given above. In order to avoid treating the class b separately we shall think of M in the equivalent presentation

$$M = \{0; (o,g,0,0); (1,b), (\alpha_1, \beta_1), \ldots, (\alpha_r, \beta_r)\} .$$

Remove the interior of an equivariant tube consisting of only principal orbits from M and call the resulting manifold-with-boundary M' . Let V be a tubular neighborhood of an E-orbit with Seifert invariants (α, β) as described in (1.7), $\alpha > 0$, $(\alpha, \beta) = 1$ but β is not necessarily normalized.
As in (1.7) define ν and ρ by

$$\nu\beta \equiv 1 \bmod \alpha \quad , \quad 0 < \nu < \alpha$$
$$\rho = (\beta\nu - 1)/\alpha .$$

Choose a cross-section on the boundary torus of M' so that the action written with complex coordinates is

$$t(z_1, z_2) = (z_1, tz_2)$$

$t \in U(1)$, $|z_1| = 1$, $|z_2| = 1$.

The action in V is described by

$$t(x,z) = (t^{\nu}x, t^{\alpha}z)$$

$|x| \leq 1$, $|z| = 1$. Define the equivariant map

$$\varphi: \partial M' \to \partial V$$

by
$$\varphi(z_1,z_2) = (z_1^{-\alpha}z_2^{\nu}, z_1^{\beta}z_2^{-\rho}).$$

Its inverse is the map F given in (1.10). Since φ has determinant -1 it is orientation reversing and it can be used to obtain an oriented manifold

$$\tilde{M} = M(\alpha,\beta) = M' \underset{\varphi}{\cup} V .$$

Let $Y_- = \tilde{M} \times I$ with $\tilde{M} = \tilde{M} \times \{0\} \subset Y_-$. Consider the unit ball in $\mathbb{C}^2$

$$D_{\nu,\alpha} = \{(z_1,z_2) \in \mathbb{C}^2 \mid |z_1|^2 + |z_2|^2 \leq 1\}$$

with the $U(1)$ action

$$t(z_1,z_2) = (t^{\nu}z_1, t^{\alpha}z_2)$$

and let $S_{\nu,\alpha} = \partial D_{\nu,\alpha}$ denote S^3 with the above action.

The map

$$\lambda(x,z) = \left(\frac{x}{\sqrt{1+x\bar{x}}}, \frac{z}{\sqrt{1+x\bar{x}}}\right)$$

defines an orientation preserving equivariant embedding $\lambda: V \to S_{\nu,\alpha}$.

Define $D^{\frac{1}{2}}_{\nu,\alpha} = \{(z_1,z_2) \in D_{\nu,\alpha} \mid |z_1|^2 + |z_2|^2 \leq \frac{1}{2}\}$

and

$$Y_+ = \overline{D_{\nu,\alpha} - D^{\frac{1}{2}}_{\nu,\alpha}} \subset D_{\nu,\alpha} .$$

Using λ sew Y_+ and Y_- together along $V \times \{1\} \subset \tilde{M} \times \{1\}$ to obtain a 4-manifold with boundary $Y = Y_- \underset{\lambda}{\cup} Y_+$ with a fixed point free S^1 action.

The boundary of Y has three components $M(\alpha,\beta) = \tilde{M} \times \{0\} \subset Y_-$,

$S^{\frac{1}{2}}_{\nu,\alpha} = S(\xi_0^{(\alpha)} \hat{\oplus} \xi_0^{(\nu)})$ and the result of sewing $\tilde{M} \times \{1\}$ and $S_{\nu,\alpha}$ together by λ. The latter is obtained by sewing the complement of V in $S_{\nu,\alpha}$ into $M' = M(\alpha,\beta) - V$. A careful analysis shows that

$$\partial Y = M(\alpha,\beta) - M(\nu,\rho) - S(\xi_0^{(\alpha)} \hat{\oplus}\, \xi_0^{(\nu)}) .$$

In order to emphasize the symmetry of the situation we let $\nu = \tilde{\alpha}$ and $\rho = \tilde{\beta}$ and write the result as:

Lemma 2. With the above notation the fixed point free S^1-manifold Y has boundary

$$\partial Y = M(\alpha,\beta) - M(\tilde{\alpha},\tilde{\beta}) - S(\xi_0^{(\alpha)} \hat{\oplus}\, \xi_0^{(\tilde{\alpha})}) .$$

Noting that $0 < \tilde{\alpha} \leq \alpha$ the above lemma gives an algorithm for representing the cobordism class of an arbitrary fixed point free S^1-manifold in terms of the generators of $\mathcal{O}_3(\infty)$ given in Theorem 1.

4.4. The α-invariant

Consider the composition of inclusion maps

$$\bar{\mathcal{O}}_*(\infty) \xrightarrow{\varphi} \mathcal{O}_*(\infty) \xrightarrow{i} \mathcal{O}_*(S^1)$$

Theorem 1. The sequence above is exact in the middle.

Corollary 2. If M is a fixed point free S^1-manifold with no isotropy group of even order, then M bounds an S^1-manifold.

Proof. By (4.2.4) $\operatorname{im}\varphi \subset \ker i$. On the other hand we have the exact sequence of (4.1)

$$\to \mathcal{O}_*(S^1) \to \bigoplus_k \psi_*^k(S^1) \overset{\partial}{\to} \mathcal{O}_*(\infty) \overset{i}{\to} \tilde{\mathcal{O}}_*(S^1) \to \cdots$$

so it is sufficient for the converse that $\ker i = \operatorname{im}\partial \subset \operatorname{im}\varphi$. This follows because an S^1-vector bundle of type (S^1) with fixed point set equal to the zero section has a natural complex structure inducing the structure of an S^1-manifold with complex normal bundle on the associated sphere bundle.

The next result is stated without proof, Ossa [1, 2.2.1].

<u>Theorem 3</u>. $\operatorname{coker}\varphi$ <u>is a 2-torsion group</u>.

Thus for every fixed point free S^1-manifold M, a suitable multiple $2^r M$ bounds an S^1-manifold. This fact will be used to define an invariant of the S^1-action on M, $\alpha(M)$ below.

Given an S^1-vectorbundle η over the compact, oriented manifold X so that the fixed point set is equal to the zero-section $X \subset \eta$, there is a canonical splitting of η into a sum of complex S^1-vectorbundles η_k, $k \geq 1$ so that $t \in S^1$ operates by complex multiplication by t^k in the fiber of η_k.

Let

$$c(\eta_k) = \prod_{j=1}^{n_k} (1 + x_j(k)) , \quad x_j(k) \text{ of degree } 2$$

be a formal factorization of the total chernclass $c(\eta_k) \in H^*(X;Q)$. Let $\mathcal{L}(X) \in H^*(X;Q)$ be the total $\mathcal{L}$ polynomial of X, Hirzebruch [2]. Define a rational function $\tilde{\alpha}(\eta) \in Q(t)$ by

$$\tilde{\alpha}(\eta) = (\mathcal{L}(X) \prod_{k>0} \prod_{j=1}^{n_k} \frac{t^k e^{2x_j(k)} + 1}{t^k e^{2x_j(k)} - 1})[X] ,$$

where $[X]$ is the fundamental class of X, $[X] \in H_*(X;Q)$.

Given a closed, oriented S^1-manifold M with fixed point set X, its normal bundle η has a canonical complex structure and therefore it induces an orientation on X from the orientation of M. If $\tau(M)$ denotes the signature of M, then a fixed point theorem in Atiyah-Singer [1,p.582] implies that

$$\tau(M) = \tilde{\alpha}(\eta) .$$

Now assume that M is an oriented fixed point free S^1-manifold. For some r we can find an oriented S^1-manifold Y so that $\partial Y = 2^r M$. Let η denote the normal bundle of the fixed point set of Y and define the rational function

$$\alpha(M) = 2^{-r}(\tau(Y) - \tilde{\alpha}(\eta)).$$

To see that $\alpha(M)$ is independent of the choice of Y one takes Y', $\partial Y' = 2^{r'} M$ and constructs

$$W = (2^{r'}Y) \underset{\partial}{\cup} (-2^r Y')$$

to obtain a closed manifold for which the Atiyah-Singer theorem applies. The additivity of the signature implies the assertion.

Remark. Ossa [1]. $\alpha(M)$ may be expressed as a polynomial in $\frac{t^k+1}{t^k-1}$, $k > 0$ with coefficients in $\mathbb{Z}[\frac{1}{2}]$.

It turns out that $\alpha(M)$ is determined up to an additive constant by the fixed point free cobordism class of M. In order to compute $\alpha(M)$ for a fixed point free 3-dimensional S^1-manifold, we first compute $\alpha(M)$ for the generators of $\mathcal{O}_3(\infty)$.

Lemma 4. Let $\eta = \xi_0^{(m)} \hat{\oplus} \xi_0^{(n)}$. Then

$$\tilde{\alpha}(\eta) = \frac{t^m+1}{t^m-1} \cdot \frac{t^n+1}{t^n-1} .$$

Let $D(\eta)$ and $S(\eta)$ be the associated disk and sphere bundles. Then clearly $\tau(D(\eta)) = 0$ and we have:

Lemma 5.

$$\alpha(S(n)) = -\frac{t^m+1}{t^m-1} \cdot \frac{t^n+1}{t^n-1} .$$

Next recall the fixed point free S^1-manifold $Y = Y(M,\alpha,\beta)$ obtained from M in (4.3) with

$$\partial Y = M(\alpha,\beta) - M(\tilde{\alpha},\tilde{\beta}) - S(\xi_0^{(\alpha)} \hat{\oplus} \xi_0^{(\tilde{\alpha})}) .$$

In order to find the relation between the α-invariants of $M(\alpha,\beta)$ and $M(\tilde{\alpha},\tilde{\beta})$ it is necessary to compute the signature of Y. Let $M = \{0;(o,g,0,0);\ (\alpha_1,\beta_1),\dots,(\alpha_{n-1},\beta_{n-1})\}$ where the (α_j,β_j) are not necessarily normalized. Direct computation gives:

Lemma 6.

$$\tau(Y) = \operatorname{sign}(\sigma + \frac{\beta}{\sigma})(\sigma + \frac{\tilde{\beta}}{\tilde{\alpha}})$$

where $\sigma = \sum_{j=1}^{n-1} \frac{\beta_j}{\alpha_j}$.

Given the relatively prime pair (α,β) of positive integers there is a unique continued fraction

$$\alpha/\beta = [a_0,a_1,\dots,a_k] = a_0 - \cfrac{1}{a_1 - \cfrac{1}{\ddots - \cfrac{1}{a_k}}}$$

with $a_i \geq 2$, as noted in (2.4). The auxiliary variables of the Euclidean algorithm are defined by $p_{-1} = 1$ $p_0 = a_0$, $p_{i+1} = a_{i+1}p_i - p_{i-1}$, $i \geq 0$. Define the rational function

$$r(\alpha,\beta) = \sum_{i=0}^{k} (1 - \frac{t^{p_i}+1}{t^{p_i}-1} \cdot \frac{t^{p_{i-1}}+1}{t^{p_{i+1}}-1}) .$$

It has the following properties

(i) $r(\alpha,-\beta) = -r(\alpha,\beta)$

(ii) $r(1,0) = 0$

(iii) if (α,β) and $(\tilde{\alpha},\tilde{\beta})$ are given so that $0 < \tilde{\alpha} \leq \alpha$ and $\alpha\tilde{\beta} - \tilde{\alpha}\beta = -1$ as above, then

$$r(\alpha,\beta) = r(\tilde{\alpha},\tilde{\beta}) + 1 - \frac{t^{\alpha}+1}{t^{\alpha}-1} \cdot \frac{t^{\tilde{\alpha}}+1}{t^{\tilde{\alpha}}-1} .$$

With this notation the α-invariant of a 3-dimensional closed, oriented S^1-manifold is computed as follows:

<u>Theorem 7</u>. <u>Let</u> $K = \{0;(o,g,0,0);(\alpha_1,\beta_1),\ldots,(\alpha_n,\beta_n)\}$. <u>Then we have</u>

$$\alpha(K) = \sum_{j=1}^{n} r(\alpha_j,\beta_j) - \text{sign}\left(\sum_{j=1}^{n} \frac{\beta_j}{\alpha_j}\right) .$$

<u>Proof</u>. We use induction assuming the statement for all $M = \{0;(o,g,0,0);(\alpha'_1,\beta'_1),\ldots,(\alpha'_m,\beta'_m)\}$ with

$m < n$ or

$m = n$ and $\alpha'_m < \alpha_n$ or

$m = n$ and $\alpha'_m = \alpha_n$ and $|\beta'_m| < |\beta_n|$.

We may assume that $\beta_n > 0$ for if $\beta_n = 0$ then the conclusion follows trivially and if $\beta_n < 0$ then we consider $-K = \{0,(o,g,0,0);(\alpha_1,-\beta_1),\ldots,(\alpha_n,-\beta_n)\}$. Let $M = \{0,(o,g,0,0);(\alpha_1,\beta_1),\ldots,(\alpha_{n-1},\beta_{n-1})\}$, $\sigma = \sum_{i=1}^{n-1} \frac{\beta_i}{\alpha_i}$ and $\alpha_n,\beta_n,\tilde{\alpha}_n,\tilde{\beta}_n$ as above. Now using the definition of α on the fixed point free S^1-manifold Y we have

$$\alpha(\partial Y) = \tau(Y)$$

$$\alpha[M(\alpha_n,\beta_n)] - \alpha[M(\tilde{\alpha}_n,\tilde{\beta}_n)] + \frac{t^{\alpha_n}+1}{t^{\alpha_n}-1} \cdot \frac{t^{\tilde{\alpha}_n}+1}{t^{\tilde{\alpha}_n}-1} = \text{sign}\left(\sigma + \frac{\beta}{\alpha}\right)\left(\sigma + \frac{\tilde{\beta}}{\tilde{\alpha}}\right) .$$

Using (iii) above and the induction hypothesis, the assertion follows from the simple identity below:

$$\operatorname{sign}(\sigma + \tfrac{\beta}{\alpha})(\sigma + \tfrac{\tilde\beta}{\tilde\alpha}) = 1 - \operatorname{sign}(\sigma + \tfrac{\beta}{\alpha}) + \operatorname{sign}(\sigma + \tfrac{\tilde\beta}{\tilde\alpha}) .$$

<u>Example 8</u>. Let us compute the α-invariant of the 3-manifold $K = \{-1;(o,3,0,0);(5,2),(5,2),(5,2),(5,2),(10,7)\}$ obtained as the neighborhood boundary of the isolated singularity at $\underline{0}$ of the surface $V = \{\underline{z} \in \mathbb{C}^3 \mid z_0^{15} + z_1^4 z_2 + z_2^7 z_1 = 0\}$ in (3.10). First we shall absorb b in the E-orbit (10,7) and write $K = \{0;(o,3,0,0);(5,2),(5,2),(5,2),(5,2),(10,-3)\}$. Next

$\frac{5}{2} = 3 - \frac{1}{2}$ and $\frac{10}{3} = 4 - \frac{1}{2-\frac{1}{2}}$. Hence

$$r(5,2) = 1 - \frac{t^3+1}{t^3-1}\,\frac{t+1}{t-1} + 1 - \frac{t^5+1}{t^5-1}\,\frac{t^3+1}{{}^3-1}$$

$$r(10,3) = 1 - \frac{t^4+1}{t^4-1}\,\frac{t+1}{t-1} + 1 - \frac{t^7+1}{t^7-1}\,\frac{t^4+1}{t^4-1} + 1 - \frac{t^{10}+1}{t^{10}-1}\,\frac{t^7+1}{t^7-1}$$

and $\displaystyle\sum_{i=1}^{5} \frac{\beta_i}{\alpha_i} = 4\cdot\frac{2}{5} + \frac{-3}{10} = \frac{13}{10}$ so

$$\alpha(K) = 4r(5,2) - r(10,3) - 1 .$$

5. Fundamental Groups

We noted in chapter 1 that only some of the Seifert manifolds admit S^1-actions but deferred the introduction of the remaining ones to this chapter. Using the terminology of Holmann [1] given in (5.1), the other Seifert manifolds are described in (5.2) and the classification theorem of Seifert [1] is proved. In (5.3) we compute the fundamental groups and use the method of Orlik-Vogt-Zieschang [1] to show that if the fundamental groups of two Seifert manifolds satisfy a condition (in which case they will be called "large"), then they are isomorphic <u>only if</u> the manifolds have the same Seifert invariants (up to orientation). This gives a homeomorphism classification of large Seifert manifolds. In (5.4) we investigate "small" Seifert manifolds (i.e. whose fundamental groups are not large) and their homeomorphism classification.

5.1 Seifert Bundles

Recall that a bundle $\xi = (X,\pi,Y)$ consists of a total space X, basis Y and continuous onto map $\pi: X \to Y$. A bundle homomorphism from $\xi' = (X',\pi',Y')$ is a pair of continuous maps $h: X \to X'$, $t: Y \to Y'$ making the diagram commutative

$$\begin{array}{ccc} X & \xrightarrow{\pi} & Y \\ h\downarrow & & \downarrow t \\ X' & \xrightarrow{\pi'} & Y' \end{array}$$

It is an isomorphism if h and t are homeomorphisms.

Following Holmann [1] we define a <u>Seifert product bundle</u> with typical fiber F as a triple $\{(F\times U)/G, p', U/G\}$ where U is a topological space, G a finite group operating on F and U (the action on U is not assumed effective) and on $F\times U$ by $g(f,u) = (gf,gu)$ and there is a commutative diagram

$$\begin{array}{ccc} F\times U & \xrightarrow{\ p\ } & U \\ \chi\big\downarrow & & \big\downarrow\tau \\ (F\times U)/G & \xrightarrow{\ p'\ } & U/G \end{array}$$

where p is projection onto the second factor, χ and τ are orbit maps of the G actions and p' is the induced map.

We call $\xi = (X,\pi,Y)$ a <u>Seifert bundle</u> with typical fiber F if it is locally isomorphic to a Seifert product bundle with typical fiber F, i.e. Y has an open cover $\{V_i,\ i\in I\}$ so that to each i we have a Seifert product bundle $\{(F\times U_i)/G_i, p_i', U_i/G_i\}$ and a commutative diagram

$$\begin{array}{ccccc} F\times U_i & & \xrightarrow{\quad p_i \quad} & & U_i \\ & \searrow^{\chi_i} & & \swarrow^{\tau_i} & \\ H_i = h_i\circ\chi_i \Big\downarrow & & (F\times U_i)/G_i \xrightarrow{\ p_i'\ } U_i/G_i & & \Big\downarrow T_i = t_i\circ\tau_i \\ & \swarrow_{h_i} & & \searrow_{t_i} & \\ \pi^{-1}(V_i) & & \xrightarrow{\quad \pi \quad} & & V_i \end{array}$$

where (h_i,t_i) give a bundle isomorphism in the lower square.

We call G a <u>structure group</u> of the Seifert bundle ξ if

(i) it contains the finite groups G_i above,

(ii) each non-empty subset of U_i, $U_i^j = T_i^{-1}(V_i\cap V_j)$ has a finite (unbranched) cover (U_{ij},σ_{ij}) where $U_{ij} = U_{ji}$ so that $T_i\circ\sigma_{ij} = T_j\circ\sigma_{ji}$,

(iii) for $V_i \cap V_j \neq \emptyset$ there is a continuous map $g_{ij}: U_{ij} \to G$ so that by defining $f_{ij}: (f,u) \to (g_{ij}(u)f,u)$ the diagram below is commutative:

$$
\begin{array}{ccc}
F \times U_{ij} & \xrightarrow{\quad f_{ij} \quad} & F \times U_{ij} \\
{\scriptstyle S_{ji}} \Big\downarrow & & \Big\downarrow {\scriptstyle S_{ij} = \chi_i \circ (i_F \times \sigma_{ij})} \\
(F \times U_j^i)/G_j & \xrightarrow{\quad h_i^{-1} \circ h_j \quad} & (F \times U_i^j)/G_i
\end{array}
$$

If the fiber F equals the structure group G acting on itself by left translations, we call it a principal Seifert bundle. The following two results of Holmann [1] will be useful later.

Theorem 1. *Let* $\xi = (X,\pi,Y)$ *be a principal Seifert bundle with structure group and fiber* G . *Assume that* X, Y *and* G *are locally compact. Then* X *is a* G*-space and the orbits of the action are the fibers of the Seifert bundle.*

Theorem 2. *Let a locally compact topological group* G *act on a locally compact space* X *so that each* $g: X \to X$ *is a proper map and all isotropy groups are finite. Then* $\xi = (X,\pi,X/G)$ *is a principal Seifert bundle with fiber and structure group* G .

Corresponding results hold in the differentiable and complex analytic cases.

Example (Holmann [1].) Let $\xi = (S^3,\pi,S^2)$ be the Seifert bundle with total space S^3 and base space S^2 given by the orbits of the S^1-action on S^3 from (1.5.1)

$$t(z_1,z_2) = (t^n z_1, t^m z_2)$$

where $(m,n) = 1$ and $S^3 = \{(z_1,z_2) \in C^2 \mid z_1\bar{z}_2 + z_2\bar{z}_2 = 1\}$. We think of the base space $S^2 = CP^1$ with homogeneous coordinates $[x_1:x_2]$. The orbit map is then given by

$$\pi(z_1,z_2) = [z_1^m : z_2^n] .$$

Consider the open sets in the base space $V_i = \{[x_1:x_2] \in CP^1 | x_i \neq 0\}$, $i = 1,2$. Let U_1 and U_2 equal the complex numbers with coordinates y_1 and y_2, and G_n and G_m the corresponding cyclic groups of order n and m. Let $\xi = \exp(2\pi i/n)$ operate on U_1 by $\xi(y_1) = \xi^{-m} y_1$ and $\eta = \exp(2\pi i/m)$ operate on U_2 by $\eta(y_2) = \eta^{-n} y_2$. Define the corresponding actions on $S^1 \times U_i$ by $\xi(x,y_1) = (\xi x, \xi^{-m} y_1)$ and $\eta(x,y_2) = (\eta x, \eta^{-n} y_2)$.

Define $T_i : U_i \to V_i$, $H_i : S^1 \times U_i \to \pi^{-1}(V_i)$ by

$$T_1(y_1) = [(1+y_1\bar{y}_1)^{\frac{n-m}{2}} : y_1^n]$$

$$T_2(y_2) = [y_2^m : (1+y_2\bar{y}_2)^{\frac{m-n}{2}}]$$

$$H_1(x,y_1) = \left(\frac{x^n}{\sqrt{1+y_1\bar{y}_1}} , \frac{x^m y_1}{\sqrt{1+y_1\bar{y}_1}}\right)$$

$$H_2(x,y_2) = \left(\frac{x^n y_2}{\sqrt{1+y_2\bar{y}_2}} , \frac{x^m}{\sqrt{1+y_2\bar{y}_2}}\right)$$

giving the required Seifert diagrams.

In order to define the action of the structure group we let $U_{12} = U_{21}$ equal the complex numbers without the origin and $U_1^2 = U_1 - \{0\}$, $U_2^1 = U_2 - \{0\}$ and define the covers $\sigma_{12} : U_{12} \to U_1^2$ by $\sigma_{12}(y) = y^m$ and $\sigma_{21} : U_{21} \to U_2^1$ by $\sigma_{21}(y) = y^{-n}|y|^{n-m}$. These maps satisfy the condition $T_2 \circ \sigma_{21} = T_1 \circ \sigma_{12}$. Finally let

$g_{21}(y) = y|y|^{-1}$, $g_{12}(y) = y^{-1}|y|$ be maps $U_{12} \to S^1$ giving rise to automorphisms f_{12} and f_{21} of $S^1 \times U_{12}$ defined by $f_{12}(x,y) = (y^{-1}|y|x,y)$, $f_{21}(x,y) = (y|y|^{-1}x,y)$ satisfying $H_2 \circ (i_{S^1} \times \sigma_{21}) \circ f_{21} = H_1 \circ (i_{S^1} \times \sigma_{12})$.

<u>Remark</u>. If we define $\hat{U}_{12} = \hat{U}_{21}$ as all complex numbers and extend the maps σ_{12} and σ_{21} to be <u>branched</u> m-fold and n-fold covers and consider the locally trivial fiber bundle $\hat{\xi}$ obtained from $S^1 \times \hat{U}_{12}$ and $S^1 \times \hat{U}_{21}$ by identifying $S^1 \times U_{12}$ and $S^1 \times U_{21}$ using f_{12}, then we see that $\hat{\xi}$ is a branched mn-fold cover of ξ branched along the two $\mathbb{E}$-orbits of ξ . In fact $\hat{\xi} = (S^3, \pi, S^2)$ is just the Hopf bundle, and the equivariant branched cover is described globally by

$$\varphi : \hat{\xi} \to \xi$$

$$\varphi(z_1, z_2) = \left(\frac{z_1^n}{\sqrt{|z_1^n|^2 + |z_2^m|^2}}, \frac{z_2^m}{\sqrt{|z_1^n|^2 + |z_2^m|^2}}\right) .$$

5.2. Seifert Manifolds

In his classical paper Seifert [1] considered the class of closed 3-manifolds satisfying the conditions

(i) the manifold decomposes into a collection of simple closed curves called fibers so that each point lies on a unique fiber,

(ii) each fiber has a tubular neighborhood V consisting of fibers so that V is a "standard fibered solid torus". The latter is the quotient of $D^2 \times S^1$ by the action of a finite cyclic group as in (1.7).

The problem is to classify all such manifolds up to fiber preserving homeomorphism. In the notation of (5.1) we have Seifert bundles $\xi = (M,\pi,B)$ where M is a closed 3-manifold, the fiber is S^1 and the structure group is all homeomorphisms of S^1. Since this group retracts onto $O(2)$ we can restate our problem as follows: Classify all Seifert bundles $\xi = (M,\pi,B)$ with total space a closed 3-manifold, fiber S^1 and structure group $O(2)$ under bundle equivalence. The first result is a consequence of (5.1.1).

Proposition 1. If the structure group reduces to $SO(2)$, then ξ is a principal Seifert bundle with typical fiber S^1. M admits an S^1-action and the classification is given by Theorem (1.10).

Considering the general case we may use the argument of (1.9) to conclude that B is a closed 2-manifold of genus g. Thus there are only finitely many open sets V_i in the cover of B with $G_i \neq 1$. A refinement of the cover enables us to collect all these in an open set at the base point of B. Outside of this set ξ is a genuine fiber bundle. The structure group $O(2)$ contains reflection of the fiber, i.e. along some curve of B (not homotopic to zero) the fiber may reverse its orientation. This gives rise to a homomorphism

$$\varphi : \pi_1(B) \to C_2$$

where C_2 is the multiplicative group of order 2, $C_2 = \{1,-1\}$ identified with the automorphism group of $\pi_1(S^1) = \mathbb{Z}$. Here $\varphi(x) = 1$ if the fiber preserves its orientation along a curve representing x and $\varphi(x) = -1$ otherwise. Select a set of generators for $\pi_1(B)$. The next result is due to Seifert [1]. We give the proof of Orlik [1], see also Orlik-Raymond [2] for generalizations.

Theorem 2. *Up to Seifert bundle equivalence there are the following six possibilities:*

o_1: B *is orientable and all generators preserve orientation so* M *is orientable and* ξ *is a principal Seifert bundle;*

o_2: B *is orientable with* $g \geq 1$ *and all generators reverse orientation so* M *is non-orientable;*

n_1: B *is non-orientable and all generators preserve orientation so* M *is non-orientable and* ξ *is a principal Seifert bundle;*

n_2: B *is non-orientable and all generators reverse orientation so* M *is orientable;*

n_3: B *is non-orientable with* $g \geq 2$ *and one generator preserves orientation while all others reverse orientation so* M *is non-orientable;*

n_4: B *is non-orientable with* $g \geq 3$ *and two generators preserve orientation while all others reverse orientation so* M *is non-orientable.*

Proof. Clearly $\varphi : \pi_1(B) \to C_2$ is determined by the values on the generators. We shall show that for an arbitrary homomorphism we can choose new generators of $\pi_1(B)$ so that the induced φ acts on the generators according to one of the maps in the theorem.

If B is orientable and φ maps all generators into +1 or all generators into -1, then there is nothing to show. Now suppose $\varphi(u_i) = -1$ and $\varphi(u_j) = 1$. By renumbering the generators we may assume $\varphi(u_1) = 1$. Let j be the smallest index so that $\varphi(u_j) = 1$. If

(i) j is even: let $v_{j-1} = u_{j-1}u_j$; $v_j = u_{j-1}$ and $v_k = u_k$ for $k \neq j-1, j$.

(ii) j is odd ($j \geq 3$) and $\varphi(u_{j+1}) = 1$: let $v_{j-1} = u_j^{-1}u_{j-1}$; $v_j = u_j^{-1}u_{j-1}u_{j-2}u_{j-1}^{-1}u_j u_{j+1}^{-1}$; $v_{j+1} = u_{j+1}u_j u_{j+1}^{-1}$ and $v_k = u_k$ for $k \neq j-1, j, j+1$;

j is odd ($j \geq 3$) and $\varphi(u_{j+1}) = -1$: let $v_j = u_j u_{j+1}$ and $v_k = u_k$ for $k \neq j$.

Repeated application of this procedure defines new generators for $\pi_1(B)$ so that φ sends every generator into -1 .

A similar argument holds if B is non-orientable. If all generators are mapped into $+1$ we have a principal bundle, n_1 . If all generators are mapped into -1 we have an orientable total space, n_2 . Now suppose that some generators preserve orientation and some reverse it. Let $\varphi(u_1) = -1$ and $\varphi(u_2) = \varphi(u_3) = \varphi(u_4) = 1$. The following change of basis reduces the number of orientation preserving generators by two:

$$v_1 = u_1u_2u_3 \; ; \qquad v_2 = u_3^{-1}u_2^{-1}u_1^{-1}u_3^{-1}u_2^{-1}u_3u_4^{-1}u_3^{-2}u_2^{-1}u_3 \; ;$$

$$v_3 = u_3^{-1}u_2u_3^2u_4 \; , \quad v_4 = u_4^{-1}u_3^{-1}u_1u_2^2u_3^2u_4^2 \; ; \quad v_i = u_i \quad \text{for } i > 4 \; .$$

Repeated application of this map gives n_3 or n_4 .

To show that the six bundle equivalence classes are indeed distinct is trivial in all cases except for n_3 and n_4 . Here we abelianize $\pi_1(B)$ and notice that the image of $u_1u_2 \ldots u_g$ is the unique element of order 2 in $H_1(B;\mathbb{Z})$. This element commutes in $\pi_1(M)$ with the homotopy class of a typical fiber for odd g only for n_3 and for even g only for n_4 .

Using the proof of the classification theorem (1.10) for 3-manifolds with S^1-action and $F \cup SE = \emptyset$, we obtain the following classification theorem of Seifert [1].

Theorem 3. Let $\xi = (M,\pi,B)$ be a Seifert bundle with typical fiber S^1, structure group $O(2)$ and total space M a closed 3-manifold. It is determined up to bundle equivalence (preserving the orientation of M or B if they have any) by the following Seifert invariants:

$$M = \{b;(\epsilon,g);(\alpha_1,\beta_1),\ldots,(\alpha_r,\beta_r)\} .$$

Here ϵ is one of o_1,o_2,n_1,n_2,n_3,n_4 denoting the weighted map of the 2-manifold B of genus g described in Theorem 2; the (α_j,β_j) are pairs of relatively prime positive integers

$0 < \beta_j < \alpha_j$ for $\epsilon = o_1,n_2$,

$0 < \beta_j \leq \alpha_j/2$ for $\epsilon = o_2,n_1,n_3,n_4$;

and b is an integer satisfying the conditions

$b \in \mathbb{Z}$ for $\epsilon = o_1,n_2$ and

$b \in \mathbb{Z}_2$ for $\epsilon = o_2,n_1,n_3,n_4$ unless $\alpha_j = 2$ for some j in which case $b = 0$.

Note that M is orientable if $\epsilon = o_1,n_2$ and a change of orientation gives the Seifert invariants

$$-M = \{-b-r;(\epsilon,g);(\alpha_1,\alpha_1-\beta_1),\ldots,(\alpha_r,\alpha_r-\beta_r)\} .$$

5.3. Fundamental Groups

The fundamental group $G = \pi_1(M)$ is generated by the "partial cross-section" $q_0,\ldots,q_r$ and $a_1,b_1,\ldots,a_g,b_g$ if B is orientable or $v_1,\dots,v_g$ if B is non-orientable and the fiber h. The relations are given by: the commuting relations of h with the other generators, the null homotopic curves in the E-orbits: $q_j^{\alpha_j} h^{\beta_j} = 1$, the relation on the "partial cross-section" $q_0 \pi_* = 1$ where $\pi_* = q_1 \dots q_r [a_1,b_1] \dots [a_g,b_g]$ if B is orientable and $\pi_* = q_1 \dots q_r v_1^2 \dots v_g^2$ if B is non-orientable, and the relation $q_0 h^b = 1$, which we eliminate by substituting $q_0 = h^{-b}$. Thus for orientable B we have

$$G = \{a_1,b_1,\dots,a_g,b_g,q_1,\dots,q_r,h \mid a_i h a_i^{-1} = h^{\epsilon_i},\ b_i h b_i^{-1} = h^{\epsilon_i},\ q_j h q_j^{-1} = h,$$
$$q_j^{\alpha_j} h^{\beta_j} = 1,\ q_1 \dots q_r [a_1,b_1] \dots [a_g,b_g] = h^b\}$$

o_1: $\epsilon_i = 1$ for all i,

o_2: $\epsilon_i = -1$ for all i;

and for non-orientable B we have

$$G = \{v_1,\dots,v_g,q_1,\dots,q_r,h \mid v_i h v_i^{-1} = h^{\epsilon_i},\ q_j h q_j^{-1} = h,\ q_j^{\alpha_j} h^{\beta_j} = 1,$$
$$q_1 \dots q_r v_1^2 \dots v_g^2 = h^b\}$$

n_1: $\epsilon_i = 1$ for all i,

n_2: $\epsilon_i = -1$ for all i,

n_3: $\epsilon_1 = 1$, $\epsilon_i = -1$ for $i > 1$,

n_4: $\epsilon_1 = \epsilon_2 = 1$, $\epsilon_i = -1$ for $i > 2$.

We call M <u>small</u> if it satisfies one of the conditions below:

(i) o_1, $g = 0$, $r \leq 2$,

(ii) o_1, $g = 0$, $r = 3$, $\frac{1}{\alpha_1} + \frac{1}{\alpha_2} + \frac{1}{\alpha_3} > 1$

(iii) $\{-2;(o_1,0);\ (2,1),(2,1),(2,1),(2,1)\}$

(iv) o_1, $g = 1$, $r = 0$,

(v) o_2, $g = 1$, $r = 0$,

(vi) n_1 , $g = 1$, $r \leq 1$,

(vii) n_2 , $g = 1$, $r \leq 1$,

(viii) n_1 , $g = 2$, $r = 0$,

(ix) n_2 , $g = 2$, $r = 0$,

(x) n_3 , $g = 2$, $r = 0$,

otherwise we call M large.

We shall assume in the remainder of this section that M is large and prove following Orlik-Vogt-Zieschang [1] that the Seifert invariants of M are determined (up to orientation) by $\pi_1(M)$. Small Seifert manifolds will be treated in the next section.

Lemma 1. The subgroup generated by h is the unique maximal cyclic normal subgroup of G and h has infinite order.

Proof. Consider the following groups:

$$C_i = \{q_i, h \mid q_i h q_i^{-1} = h,\ q_i^{\alpha_i} h^{\beta_i} = 1\}$$

$$D_i = \{a_i, b_i, h \mid a_i h a_i^{-1} = h^{\varepsilon_i},\ b_i h b_i^{-1} = h^{\varepsilon_i}\}$$

$$E_i = \{v_i, h \mid v_i h v_i^{-1} = h^{\varepsilon_i}\} .$$

The subgroup generated by h is infinite cyclic and normal in each of these groups. We form the iterated amalgamated free product along (h) to obtain G as follows:

(i) for orientable B and $r > 3$ we take

$$C_1 \underset{(h)}{*} C_2$$

and note that h and q_1q_2 form a free abelian subgroup of rank 2. Taking

$$C_3 \underset{(h)}{*} C_4 \underset{(h)}{*} \cdots \underset{(h)}{*} C_r \underset{(h)}{*} D_1 \underset{(h)}{*} \cdots \underset{(h)}{*} D_g$$

we find that h and $(q_3 \ldots q_r \Pi [a_i, b_i] h^{-b})^{-1}$ also form a free abelian group of rank 2 so we can amalgamate along these subgroups. A similar argument shows the assertion for all classes except for o_1, $g = 0$, $r = 3$, $\frac{1}{\alpha_1} + \frac{1}{\alpha_2} + \frac{1}{\alpha_3} \leq 1$, o_1, $g = 1$, $r = 1$ and o_2, $g = 1$, $r = 1$, where there are not enough "parts". For these cases we note that the quotient group $G/(h)$ is a planar discontinuous group and has no cyclic normal subgroup,

(ii) for non-orientable B the above argument works for all large Seifert manifolds. This completes the proof.

We should remark here the following well known fact.

Proposition 2. *Let K be a closed 3-manifold. If K is orientable, let $K' = K$ if not, let K' equal the orientable double cover of K. Suppose that $\pi_1(K')$ is infinite, not cyclic and not a free product. Then K and K' are aspherical and $\pi_1(K)$ has no element of finite order.*

From this follows immediately:

Proposition 3. *A large Seifert manifold M is a $K(G,1)$ space.*

We shall see later that it follows from Waldhausen [1] that they are also irreducible 3-manifolds.

Given the planar discontinuous group D defined by $\{\bar{q}_1, \ldots, \bar{q}_r, \bar{a}_1, \bar{b}_1, \ldots, \bar{a}_g, \bar{b}_g \mid \bar{q}_j^{\alpha_j} = 1,\ \bar{q}_1 \ldots \bar{q}_r [\bar{a}_1, \bar{b}_1] \ldots [\bar{a}_g, \bar{b}_g] = 1\}$ or $\{\bar{q}_1, \ldots, \bar{q}_r, \bar{v}_1, \ldots, \bar{v}_g \mid \bar{q}_j^{\alpha_j} = 1,\ \bar{q}_1 \ldots \bar{q}_r \bar{v}_1^2 \ldots \bar{v}_g^2 = 1\}$

We define free groups $\hat{D}$ with generators $\bar{Q}_1,\ldots,\bar{Q}_r,\bar{A}_1,\bar{B}_1,\ldots$ $\ldots,\bar{A}_g,\bar{B}_g$ or $\bar{Q}_1,\ldots,\bar{Q}_r, \bar{V}_1,\ldots,\bar{V}_g$ and words in these groups $\bar{\pi}_* = \bar{Q}_1\ldots\bar{Q}_r[\bar{A}_1,\bar{B}_1]\ldots[\bar{A}_g,\bar{B}_g]$ or $\bar{\pi}_* = \bar{Q}_1\ldots\bar{Q}_r\bar{V}_1^2\ldots$ $\ldots\bar{V}_g^2$. Define a homomorphism $\hat{D} \to D$ by mapping capital letters into lower case letters. Let $\omega(\bar{x}) = \omega(\bar{X}) = 1$ if we have an orientable fundamental domain and $\omega(\bar{x}) = \omega(\bar{X}) = \pm 1$ according to whether the $\bar{v}_i$ (or $\bar{V}_i$) occur an even or odd number of times in $\bar{x}$ (or $\bar{X}$).

Define the group $\hat{G}$ as either

$\{Q_1,\ldots,Q_r,A_1,B_1,\ldots,A_g,B_g,H \mid A_iHA_i^{-1} = H^{\epsilon_i}, B_iHB_i^{-1} = H^{\epsilon_i}, Q_jHQ_j^{-1} = H\}$ or $\{Q_1,\ldots,Q_r,V_1,\ldots,V_g \mid V_iHV_i^{-1} = H^{\epsilon_i}, Q_jHQ_j^{-1} = H\}$ where the ϵ_i are the same as in the definition of G. Let Π_* be as above (without bars) and define the homomorphism $\hat{G} \to G$ by sending capital letters to lower case letters. The map ω is defined as above for G and $\hat{G}$, i.e. $\omega(x) = \omega(X) = 1$ for $x \in G$ and $X \in \hat{G}$ if B is orientable and $\omega(x) = \pm 1$ $(\omega(X) = \pm 1)$ according to the parity of the number of times v_i (V_i) occur in x (X).

The next result is due to Zieschang [1].

Lemma 4. *Every automorphism A of D is induced by an automorphism $\hat{A}$ of $\hat{D}$ with the property that:*

$$\hat{A}(\bar{Q}_i) = \bar{M}_i\, \bar{Q}_{\nu_i}^{\zeta_i}\, \bar{M}_i^{-1}$$

$$\hat{A}(\bar{\Pi}_*) = \bar{M}\, \bar{\Pi}_*^{\zeta}\, \bar{M}^{-1}$$

where $\begin{pmatrix} 1 \ldots r \\ \nu_1 \ldots \nu_r \end{pmatrix}$ *is a permutation with* $\alpha_{\nu_i} = \alpha_i$ *and* $\omega(\bar{M}_i)\zeta_i = \omega(\bar{M})\zeta = \zeta = \pm 1$.

This allows us to prove the following:

Theorem 5. *Let M and M' be large Seifert manifolds and $I: G' \to G$ an isomorphism. Then we have*

$$I(q_i') = h^{\lambda_i} m_i \, q_{\nu_i}^{\zeta_i} \, m_i^{-1}$$

where $\begin{pmatrix} 1 \dots r \\ \nu_1 \dots \nu_r \end{pmatrix}$ *is a permutation and* $\omega(m_i)\zeta_i = \rho = \pm 1$. *The map* I *is induced by an isomorphism of the groups* $\hat{I}: \hat{G}' \to \hat{G}$ *where*

$$\hat{I}(Q_i') = H^{\lambda_i} M_i \, Q_{\nu_i}^{\zeta_i} \, M_i^{-1}$$

$$\hat{I}(\Pi_*') = H^{\lambda} M \, \Pi_*^{\zeta} \, M^{-1}$$

and $\omega(M)\zeta = \rho$. *Moreover* $\lambda = \sum_{i=1}^{r} \lambda_i + 2\sigma$ *where* $\sigma = 0$ *for* $\epsilon = o_1$ *or* n_2 .

Proof. Since (h) and (h') generate characteristic subgroups, the isomorphism I induces a commutative diagram:

$$\begin{array}{ccccccccc} 0 & \longrightarrow & (h') & \longrightarrow & G' & \longrightarrow & D' & \longrightarrow & 1 \\ & & \approx\downarrow I_1 & & \approx\downarrow I & & \approx\downarrow I_0 & & \\ 0 & \longrightarrow & (h) & \longrightarrow & G & \longrightarrow & D & \longrightarrow & 1 \end{array}$$

Next define an inclusion map $\hat{\varphi}: \hat{D} \to \hat{G}$ by $\bar{Q}_i \to Q_i$, $\bar{A}_i \to A_i$, $\bar{B}_i \to B_i$, $\bar{V}_i \to V_i$ and consider the diagram below where $\hat{I}_0$ is defined to induce I_0 by lemma 4.

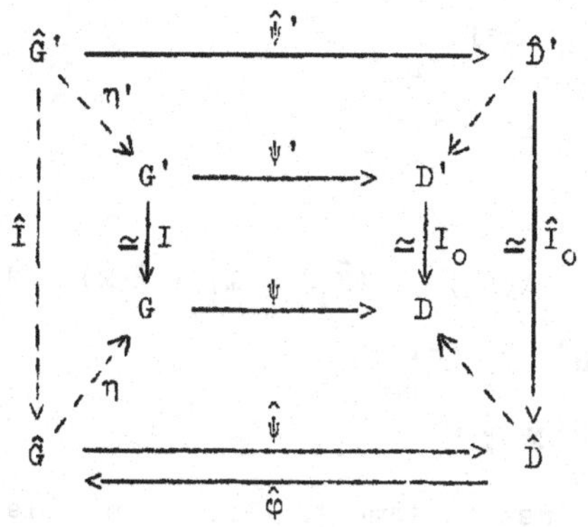

Considering the solid arrows only this diagram is commutative. We want to lift the isomorphism I to an isomorphism $\hat{I}$ of the "^" groups. Let η and η' send capital letters to lower case letters. We can construct generators for $\hat{G}'$ from generators of $\hat{G}$ using the composition $\hat{J} = \hat{\varphi}\hat{I}_o\hat{\psi}'$. In order to make the whole diagram commute (apart from $\hat{\varphi}$), we note that the difference between $I\eta'$ and $\eta\hat{J}$ lies in the kernel of ψ, (h). Now suppose that X' is a generator of $\hat{G}'$ and

$$h^{\lambda(X')}\eta\,\hat{J}(X') = I\eta'(X') .$$

Define $\hat{I}$ by

$$\hat{I}(X') = H^{\lambda(X')}\hat{J}(X')$$

$$\hat{I}(H') = H^{\delta}$$

where $I(h') = h^{\delta}$ and $\delta = \pm 1$ from I_1 above.
This makes the diagram

$$\begin{array}{ccccccccc} 0 & \longrightarrow & (H') & \longrightarrow & \hat{G}' & \longrightarrow & \hat{D}' & \longrightarrow & 1 \\ & & \downarrow{\scriptstyle \hat{I}_1} & & \downarrow{\scriptstyle \hat{I}} & & \downarrow{\scriptstyle \hat{I}_o} & & \\ 0 & \longrightarrow & (H) & \longrightarrow & \hat{G} & \longrightarrow & \hat{D} & \longrightarrow & 1 \end{array}$$

commutative so $\hat{I}$ is an isomorphism. It follows from lemma 4 that

$$\hat{I}_o(\bar{Q}_i') = \bar{M}_i\,\bar{Q}_{\nu_i}^{\zeta_i}\,\bar{M}_i^{-1}$$

$$\hat{I}_o(\bar{\Pi}_*') = \bar{M}\,\bar{\Pi}_*^{\zeta}\,\bar{M}^{-1}$$

with $\omega(\bar{M}_i)\zeta_i = \omega(\bar{M})\zeta = \rho$.
Letting $\lambda_i = \lambda(Q_i')$, $\lambda = \lambda(\Pi_*')$, $\hat{\varphi}(\bar{M}_i) = M_i$, $\hat{\varphi}(\bar{M}) = M$ we have

$$\hat{I}(Q_i') = H^{\lambda_i}\,M_i\,Q_{\nu_i}^{\zeta_i}\,M_i^{-1}$$

$$\hat{I}(\Pi_*') = H^{\lambda}\,M\,\Pi_*^{\zeta}\,M^{-1} .$$

It remains to prove the last statement. For orientable B we have

$$H^{-\lambda}\hat{I}(\Pi_*^!) = \hat{J}(\Pi_*^!) = \hat{J}(Q_1^!)\ldots\hat{J}(Q_r^!)[\hat{J}(A_1^!),\hat{J}(B_1^!)]\ldots[\hat{J}(A_g^!),\hat{J}(B_g^!)] =$$

$$H^{-\lambda_1}\hat{I}(Q_1^!)\ldots H^{-\lambda_r}\hat{I}(Q_r^!)[H^{-\lambda(A_1^!)}\hat{I}(A_1^!),H^{-\lambda(B_1^!)}\hat{I}(B_1^!)]\ldots$$

$$[H^{-\lambda(A_g^!)}\hat{I}(A_g^!),H^{-\lambda(B_g^!)}\hat{I}(B_g^!)] .$$

If $A_i^!$ and $B_i^!$ commute with H' then so do $\hat{I}(A_i^!)$ and $\hat{I}(B_i^!)$ and their commutator equals $[\hat{I}(A_i^!),\hat{I}(B_i^!)]$, thus $\lambda = \sum_{i=1}^{r}\lambda_i$. If $A_i^!$ and $B_i^!$ anticommute with H' then the corresponding commutator equals

$$H^{-2\lambda(A_i^!)-2\lambda(B_i^!)}[\hat{I}(A_i^!),\hat{I}(B_i^!)] \quad \text{so} \quad \lambda = \sum_{i=1}^{r}\lambda_i + 2\sigma .$$

For non-orientable B a similar argument works.

This leads us to the following homeomorphism classification theorem for large Seifert manifolds.

Theorem 6. Let M and M' be large Seifert manifolds. The following statements are equivalent:

(i) M and M' are equivalent Seifert bundles (possibly after reversing the orientation of one),

(ii) M and M' are homeomorphic,

(iii) M and M' have isomorphic fundamental groups.

Proof. Clearly (i) ==> (ii) ==> (iii). Assume that we have an isomorphism $I : G' \to G$. Assume moreover that the permutation of theorem 5 is the identity. By lemma 1 we have an induced isomorphism $I_0 : G'/(h') \to G/(h)$ between non-euclidean crystallographic groups. This shows that $B' = B$, $g' = g$, $r' = r$ and $\alpha_i^! = \alpha_i$. Also by lemma 1 $I(h') = h^\delta$ with $\delta = \pm 1$. Applying

I to the relation $q_i'^{\alpha_i} h'^{\beta_i'} = 1$ according to theorem 5 gives

$$1 = (h^{\lambda_i} m_i q_i^{\zeta_i} m_i^{-1})^{\alpha_i} h^{\beta_i'\delta} = m_i q_i^{\alpha_i\zeta_i} m_i^{-1} h^{\lambda_i\alpha_i+\delta\beta_i'} =$$

$$m_i h^{-\beta_i\zeta_i} m_i^{-1} h^{\lambda_i\alpha_i+\delta\beta_i'} = h^{-\epsilon(m_i)\zeta_i\beta_i+\lambda_i\alpha_i+\delta\beta_i'}$$

where for $x \in G$ we let $\epsilon(x) = \pm 1$ according to whether x commutes with h or anticommutes with h. Since h has infinite order

$$-\epsilon(m_i)\zeta_i\beta_i + \lambda_i\alpha_i + \delta\beta_i' = 0 .$$

For o_1 and n_2 we have $\epsilon(m_i) = \omega(m_i)$ so $\epsilon(m_i)\zeta_i = \omega(m_i)\zeta_i = \rho$. Thus

$$\beta_i = \rho\delta\beta_i' + \rho\lambda_i\alpha_i$$

and if $\rho\delta = 1$ then the condition $0 < \beta_i < \alpha_i$ implies that $\lambda_i = 0$ while if $\rho\delta = -1$ we get $\delta\lambda_i = -1$. Substituting these values we have $\beta_i = \beta_i'$ or $\beta_i = \alpha_i - \beta_i'$ for all i. For the other classes the condition $0 < \beta_i \leq \alpha_i/2$ implies that $\beta_i = \beta_i'$ and $\lambda_i = 0$ for all i.

Finally we need a similar computation for b:

$$1 = I(\pi_*' h'^{-b'}) = h^{\lambda} m \pi_*^{\zeta} m^{-1} h^{-\delta b'} =$$

$$h^{\lambda} m h^{\zeta b} m^{-1} h^{-\delta b'} = h^{\lambda+\epsilon(m)\zeta b-\delta b'}$$

and since h has infinite order

$$\lambda + \epsilon(m)\zeta b - \delta b' = 0 .$$

For o_1 and n_2 we have $\epsilon(m) = \omega(m)$, $\omega(m)\zeta = \rho$ and $\lambda = \sum_{i=1}^{r} \lambda_r$ so

$$\sum_{i=1}^{r} \lambda_i + \rho b - \delta b' = 0$$

if $\rho\delta = 1$ then $\lambda_i = 0$ and $b = b'$; if $\rho\delta = -1$ then $\delta\lambda_i = 1$ and $b = -b'-r$ as required.

For the other classes $\lambda_i = 0$ and $\lambda = 2\sigma$ but $b, b' \in \mathbb{Z}_2$ so $b = b'$. This completes the proof.

5.4 Small Seifert Manifolds

This section is based on Orlik-Raymond [2].

(i) The manifolds o_1, $g = 0$, $r \leq 2$ (lens spaces). Since these manifolds all admit S^1-actions we can use the equivariant method of chapter 2 to identify them. The manifold $L(b,1) = \{b;(o_1,0)\}$ was discussed there. The standard orientation gives $S^3 = L(-1,0) = L(1,1)$ and we note that $L(0,1) = S^2 \times S^1$.

The manifold $\{b;(o_1,0);(\alpha,\beta)\}$ is identified similarly. By lemma (2.2.3) it is the boundary of the linear plumbing according to the graph

$$-b-1 \quad -b_1 \quad -b_2 \quad \cdots \quad -b_s$$

where $\frac{\alpha}{\alpha-\beta} = [b_1,\ldots,b_s]$. According to lemma (2.2.1) the result of this linear plumbing is $L(p,q)$ where

$$\frac{p}{q} = [b+1,b_1,\ldots,b_s] = b+1 - \frac{1}{\frac{\alpha}{\alpha-\beta}} = \frac{\alpha(b+1)-(\alpha-\beta)}{\alpha} = \frac{b\alpha+\beta}{\alpha}$$

so we see that $\{b;(o_1,0);(\alpha,\beta)\} = L(b\alpha+\beta,\alpha)$.

For $r = 2$ we apply the same argument: $\{b;(o_1,0);(\alpha_1,\beta_1),(\alpha_2,\beta_2)\}$ is the boundary of the equivariant linear plumbing

$$-b_{1,s_1} \quad -b_{1,s_1-1} \quad \cdots \quad -b_{1,1} \quad -b-2 \quad -b_{2,1} \quad \cdots \quad -b_{2,s_2}$$

where $\frac{\alpha_1}{\alpha_1-\beta_1} = [b_{1,1},\ldots,b_{1,s_1}]$ and $\frac{\alpha_2}{\alpha_2-\beta_2} = [b_{2,1},\ldots,b_{2,s_2}]$.

It is $L(p,q)$ with

$$\frac{p}{q} = [b_{1,s_1},\ldots,b_{1,1},b+2,b_{2,1},\ldots,b_{2,s_2}] .$$

First we note that the result of a reverse plumbing

$$-b_s \quad \cdots \quad -b_1$$

is determined from the product of matrices

$$\begin{pmatrix}-1 & 0\\ b_1 & 1\end{pmatrix}\begin{pmatrix}0 & 1\\ 1 & 0\end{pmatrix}\cdots\begin{pmatrix}0 & 1\\ 1 & 0\end{pmatrix}\begin{pmatrix}-1 & 0\\ b_s & 1\end{pmatrix} = \begin{pmatrix}-\bar{p}_{s-1} & -\bar{p}'_{s,1}\\ \bar{p}_s & \bar{p}'_s\end{pmatrix}$$

and by induction

$$\bar{p}_s = p_s\ ,\quad \bar{p}'_s = p_{s-1}\ ,\quad \bar{p}_{s-1} = p'_s\ ,\quad \bar{p}'_{s-1} = p'_{s-1}\ ,$$

Thus we have for the determination of $L(p,q)$ using (2.2.3):

$$\begin{pmatrix}\nu_2 & -\rho_2\\ \alpha_2 & \alpha_2-\beta_2\end{pmatrix}\begin{pmatrix}0 & 1\\ 1 & 0\end{pmatrix}\begin{pmatrix}-1 & 0\\ b+2 & 1\end{pmatrix}\begin{pmatrix}0 & 1\\ 1 & 0\end{pmatrix}\begin{pmatrix}\beta_1-\alpha_1 & -\rho_1\\ \alpha_1 & -\nu_1\end{pmatrix} =$$

$$= \begin{pmatrix} * & * \\ b\alpha_1\alpha_2+\alpha_1\beta_2+\alpha_2\beta_1 & m\alpha_2-n\beta_2\end{pmatrix}$$

where $m = -b\nu_1-\nu_1-\rho_1$, $n = -\nu_1$ satisfy the condition

$$m\alpha_1 - n(b\alpha_1+\beta_1) = 1\ .$$

The manifold is $L(p,q)$ with $p = b\alpha_1\alpha_2+\alpha_1\beta_2+\alpha_2\beta_1$ and $q = m\alpha_2-n\beta_2$.

The mutual homeomorphism classification of these manifolds is given by the well-known classification of lens spaces: $L(p,q)$ and $L(p',q')$ are homeomorphic if and only if $|p| = |p'|$ and $q\pm q' \equiv 0 \bmod p$ or $q\cdot q' \equiv \pm 1 \pmod p$. The fact that they are not homeomorphic to any other Seifert manifold will follow once we have proved that they are the only ones with finite cyclic fundamental groups.

(ii) <u>The manifolds</u> $o_1,\ g = 0\ ,\ r = 3\ ,\ \frac{1}{\alpha_1}+\frac{1}{\alpha_2}+\frac{1}{\alpha_3} > 1$. There are only four possible sets of α_i satisfying these conditions called the "platonic triples": $(2,2,\alpha_3)$, $(2,3,3)$, $(2,3,4)$ and $(2,3,5)$. They have finite, non-abelian fundamental groups and will be discussed in detail in the next chapter where we shall also show that those with $(2,2,\alpha_3)$ called "prism manifolds" are homeomorphic to manifolds n_2, $g = 1$, $r \leq 1$. Note that (h) is in the center of $\pi_1(M)$ and

$$\pi_1(M)/(h) = \{q_1, q_2, q_3 \mid q_1 q_2 q_3 = q_1^{\alpha_1} = q_2^{\alpha_2} = q_3^{\alpha_3} = 1\}$$

has no center so (h) is the whole center and the α_j are invariants of $\pi_1(M)$. The order of $H_1(M;\mathbb{Z})$

$$p = |b\alpha_1\alpha_2\alpha_3 + \beta_1\alpha_2\alpha_3 + \alpha_1\beta_2\alpha_3 + \alpha_1\alpha_2\beta_3|$$

is sufficient to distinguish the manifolds with given $(\alpha_1,\alpha_2,\alpha_3)$ up to orientation. Since we shall see that the only other Seifert manifolds with finite fundamental groups are the lens spaces and the prism manifolds, their homeomorphism classification is completed.

(iii) <u>The manifold</u> $M = \{-2; o_1, 0); (2,1),(2,1),(2,1),(2,1)\}$ is homeomorphic to $M' = \{0; (n_2,2)\}$. This is seen by noting that the orientable S^1 bundle over the Moebius band is homeomorphic to the manifold obtained by sewing two E-orbits of type (2,1) into a fibered solid torus. Doubling the former by an orientation reversing homeomorphism gives M'. Doubling the latter by an orientation reversing homeomorphism gives $\{0; (o_1,0),(2,1),(2,1),(2,-1),(2,-1)\} = M$. We shall see in chapter 7 that M fibers over S^1 with fiber the torus and the self-homeomorphism of the fiber is of order 2. It turns out that M is a flat Riemannian manifold doubly covered by $S^1 \times S^1 \times S^1$ and the covering can be made equivariant with respect to the S^1 action on M, see chapter 8.

The other small Seifert manifolds are easily seen not to be

homeomorphic to each other or any of the large ones with the exceptions mentioned below, compare Orlik-Raymond [2]. We shall briefly mention their special properties and return to them in chapter 7.

(iv) The manifolds $\{b; (o_1,1)\}$ are torus bundles over S^1.

(v) The manifolds $\{b; (o_2,1)\}$ are Klein bottle bundles over S^1.

(vi) The manifolds n_1, $g = 1$, $r \leq 1$ give rise to the different S^1 actions on $P^2 \times S^1$ and N, the non-orientable S^2-bundle over S^1.

(vii) The manifolds n_2, $g = 1$, $r \leq 1$. Here $M = \{0;(n_2,1)\}$ is seen as the result of taking $S^2 \times I$ fibered by intervals $p \times I$ and collapsing each boundary component by the antipodal map. The sphere $S^2 \times \{\frac{1}{2}\}$ decomposes M into a connected sum of two real projective spaces, $M = \mathbb{R}P^3 \# \mathbb{R}P^3$. The other manifolds are homeomorphic to the prism manifolds of (ii) and will be treated in detail in the next chapter as orbit spaces of finite groups acting freely on S^3.

(viii) The manifolds $\{b; (n_1,2)\}$ are the same two Klein bottle bundles as under (v).

(ix) The manifolds $\{b; (n_2,2)\}$ are torus bundles over S^1 distinct from (iv).

(x) The manifolds $\{b; (n_3,2)\}$ are the "other two" Klein bottle bundles over S^1 not obtained in (v) and (viii).

6. Free Actions of Finite Groups on S^3

There has been no significant progress in the problem of finding all 3-manifolds with finite fundamental group since the results of H. Hopf [1] and Seifert and Threlfall [1] determining orthogonal actions on S^3. These articles are somewhat difficult to read and the object of this chapter is to present old knowledge with new terminology. The basic theorem of section 1 is that if G is a finite subgroup of $SO(4)$ acting freely on S^3, then there is an action of S^1 on S^3 commuting with G so that the orbit space S^3/G is again an S^1-manifold. Thus the orbit spaces of orthogonal actions are S^1-manifolds with finite fundamental groups. These are discussed in section 2. In section 3 we list following Milnor [2] the groups that satisfy the algebraic conditions for an action but do not act orthogonally.

The intriguing fact remains that if one could find a 3-manifold with finite fundamental group not homeomorphic to one listed above, then either it would be the orbit space of a non-orthogonal action on S^3 or its universal cover would provide a counterexample to the 3-dimensional Poincaré conjecture.

6.1. Orthogonal Actions on S^3

In order to understand the structure of finite subgroups of $SO(4)$ that can act freely on S^3, we shall decompose $SO(4)$. It is useful to think of $SO(4)$ both as a group of orthogonal transformations of R^4 and as a matrix group of 4×4 real orthonormal matrices. It is clear that the maximal torus of $SO(4)$ is $T^2 = SO(2) \times SO(2)$ and the center is generated by the identity map e and the antipodal map $a = -e$. Let $C = \{e,a\}$ denote the center of $SO(4)$.

Lemma 1. The following sequence is exact:

$$1 \to C \xrightarrow{i} SO(4) \xrightarrow{p} SO(3) \times SO(3) \to 1 .$$

Proof. From Lie group theory we have that Spin(4)/center = SO(4)/C = Ad Spin(4) = Ad(Spin(3) × Spin(3)) = Spin(3)/center × Spin(3)/center = SO(3) × SO(3) .

In order to gain geometric insight we shall now give a direct proof. Consider the maximal torus T^2 given by the matrices

$$\begin{pmatrix} \cos\varphi & -\sin\varphi & 0 & 0 \\ \sin\varphi & \cos\varphi & 0 & 0 \\ 0 & 0 & \cos\psi & -\sin\psi \\ 0 & 0 & \sin\psi & \cos\psi \end{pmatrix} \qquad \begin{matrix} 0 \le \varphi < 2\pi \\ 0 \le \psi < 2\pi . \end{matrix}$$

The subgroup generated by all 1-dimensional circles $\varphi = \psi$ is called right rotations, R . The subgroup generated by $\varphi \equiv -\psi$ mod 2π is called left rotations, L . Note that $R \cap L = C$ and abstractly $R \approx L \approx S^3$. Every element $g \in SO(4)$ is decomposed into a right and left rotation but this decomposition is only defined modulo a . Moreover, every right rotation commutes with every left rotation and vica versa. Specifically, if we choose coordinates so that g is given by the matrix above, then for some right rotation by χ_r and left rotation by χ_l we have

$$\varphi = \chi_r + \chi_l + 2k\pi$$
$$\psi = \chi_r - \chi_l + 2k'\pi$$

and hence

$$\chi_r = \tfrac{1}{2}(\varphi+\psi) + (k+k')\pi$$
$$\chi_l = \tfrac{1}{2}(\varphi-\psi) + (k-k')\pi$$

are the possible choices of angles for right and left rotations. Thus g can be decomposed into two pairs (χ_r, χ_l) and $(\chi_r+\pi, \chi_l+\pi)$ differing by the antipodal map. In order to eliminate this indeterminacy we construct double covers $p_r : R \to SO(3)$ and $p_l : L \to SO(3)$ as follows:

Given a vector $\underline{v}$ in R^4 and a right rotation r by the angle χ_r, there is a unique plane through $\underline{v}$ rotated in itself by r. There is also a unique left rotation l rotating the same plane by $\chi_l = -\chi_r$ so that the rotation rl leaves $\underline{v}$ fixed. It rotates the R^3 perpendicular to $\underline{v}$ by an angle $\chi_r' = \chi_r - \chi_l = 2\chi_r$. The same construction applies for left rotations.

Thus if $g \in SO(4)$ is determined in a suitable coordinate system by the angles (φ, ψ), then its image in $SO(3) \times SO(3)$ may be identified by two R^3 rotations (χ_r', χ_l') fixing a given vector where

$$\chi_r' \equiv \varphi + \psi \quad , \quad \chi_l' \equiv \varphi - \psi \quad (\text{mod } 2\pi) .$$

<u>Lemma 2</u>. <u>If</u>

$$\chi_r' \equiv \chi_l' \equiv \pi \quad (\text{mod } 2\pi)$$

<u>then both</u> g <u>and</u> ag <u>have fixed points on</u> S^3 . <u>If</u>

$$\chi_r' \equiv \pm\chi_l' \quad (\text{mod } 2\pi)$$

<u>then either</u> g <u>or</u> ag <u>has fixed points on</u> S^3 . <u>If neither congruence holds then both</u> g <u>and</u> ag <u>are free on</u> S^3 .

<u>Proof</u>. Recall that $\varphi \equiv \chi_r + \chi_l$ (mod 2π) and $\psi \equiv \chi_r - \chi_l$ (mod 2π) so g has fixed points on S^3 if and only if at least one of these angles is zero so $\chi_r \pm \chi_l \equiv 0$ (mod 2π) . From the

relations $\chi'_r \equiv \pm 2\chi_r$, $\chi'_1 \equiv \pm 2\chi_1 \pmod{2\pi}$ we obtain the required formuli. The converse is a similar computation.

Let $G \subset SO(4)$ be a finite subgroup acting freely on S^3. Let $H = p(G)$ and $H_1 = pr_1 H \subset SO(3)$, $H_2 = pr_2 H \subset SO(3)$. Then clearly $H \subset H_1 \times H_2$ but H itself is not necessarily a direct product of subgroups.

The finite subgroups of $SO(3)$ were first found by F. Klein. They are the

<u>cyclic group</u> C_n of order n , $C_n = \{x \mid x^n = 1\}$;

<u>dihedral group</u> D_{2n} of order $2n$, the group of space symmetries of a regular plane n-gon generated by rotations and a reflection

$$D_{2n} = \{x,y \mid x^2 = y^n = (xy)^2 = 1\} ;$$

<u>tetrahedral group</u> T of order 12, the group of symmetries of a regular tetrahedron,

$$T = \{x,y \mid x^2 = (xy)^3 = y^3 = 1\} ;$$

<u>octahedral group</u> O of order 24, the group of symmetries of a regular octahedron or , equivalently the cube

$$O = \{x,y \mid x^2 = (xy)^3 = y^4 = 1\} ;$$

<u>icosahedral group</u> I of order 60, the group of symmetries of a regular icosahedron or , equivalently the dodecahedron

$$I = \{x,y \mid x^2 = (xy)^3 = y^5 = 1\} .$$

<u>Lemma 3</u>. <u>Every finite subgroup of</u> $SO(3)$ <u>is one of the above</u>.

Proof. (Wolf [1]) If G is a finite subgroup of $SO(3)$ and $g \in G$ $g \neq 1$, then g is a rotation by an angle θ_g about a line L_g through the origin. Let P_g be the intersection of L_g with the unit sphere S^2 consisting of the two "poles" $P_g = \{p_g, p_g'\}$ which are the only fixed points of g on S^2. We call two points $x, y \in S^2$ G-equivalent if $gx = y$ for some $g \in G$. Let $\{C_1, \ldots, C_q\}$ be the equivalence classes of poles for all non-trivial elements of G. If p is a pole, let G_p be the subgroup preserving p : $G_p = 1 \cup \{g \in G - 1 \mid p \in P_g\}$. Let p belong to the class C_i and enumerate C_i as $\{g_1 p, g_2 p, \ldots g_{r_i} p\}$ with $g_1 = 1$ and the g_i a system of representatives of the cosets of G_p in G. In particular $G_{g_i p} = g_i G_p g_i^{-1}$ exhaust all the conjugates of G_p in G and the $G_{g_i p}$ all have the same order n_i. If N is the order of G then $N = r_i n_i$.

Note that G has $N - 1$ non-trivial elements and each one has 2 poles. Since exactly $n_i - 1$ non-trivial elements of G preserve a pole $p \in C_i$ we have the identity

$$2(N-1) = \sum_{i=1}^{q} r_i(n_i - 1)$$

so

$$2\left(1-\frac{1}{N}\right) = \sum_{i=1}^{q} \left(1-\frac{1}{n_i}\right).$$

Since $N \geq n_i \geq 2$ we see that q is 2 or 3 and one of the following must hold:

(i) $q = 2$, $n_1 = n_2 = N > 1$

(ii) $q = 3$, $2 = n_1 \leq n_2 \leq 3$ $n_2 \leq n_3$ with the possibilities

a) $n_1 = n_2 = 2$, $N = 2n_3 \geq 4$,

b) $n_1 = 2$, $n_2 = n_3 = 3$, $N = 12$,

c) $n_1 = 2$, $n_2 = 3$, $n_3 = 4$, $N = 24$,

d) $n_1 = 2$, $n_2 = 3$, $n_3 = 5$, $N = 60$.

It is now a simple geometric argument to show that these cases indeed correspond to the already listed groups.

We can now combine lemmas 2 and 3 noting that D_{2n}, T, O and I have elements of even order and go through the possible subgroups of $H_1 \times H_2$ to obtain:

Lemma 4. At least one of H_1 and H_2 is cyclic.

This enables us to prove the main theorem of this section due to Seifert and Threlfall [1].

Theorem 5. Let G be a finite subgroup of $SO(4)$ acting freely on S^3. Then there is an S^1-action on S^3 so that the action of G is equivariant and the orbit space S^3/G is again an S^1-manifold.

Proof. We may assume that H_1 is cyclic. Since $R \approx S^3$, its preimage $G_r = p_r^{-1}(H_1)$ is cyclic and we can embed it in a circle subgroup Σ of R. Note that this is not true of every cyclic subgroup of $SO(4)$. Since every element of G decomposes into a left and a right rotation and the left rotations commute with Σ while the right rotations are contained in Σ we see that G is equivariant with respact to Σ.

It is easy to see by direct argument that the converse is also true, i.e. every S^1-manifold with finite fundamental group is the orbit space of a free orthogonal action of a finite group on S^3. We shall list the groups and the orbit spaces in the next section.

6.2 Groups and Orbit Spaces

We proved in (6.1) that if G is a finite subgroup of $SO(4)$ acting freely on S^3 and $H_1 \subset SO(3)$, $H_2 \subset SO(3)$ are the projections of G, then either H_1 or H_2 is cyclic. Assume that H_1 is cyclic of order m . Before we list the possible groups note that if G has even order, then $a \in G$ and $G/C \approx H$ so G is a C_2 central extension of H . Writing $H = \{e,h_1,\ldots,h_k\}$ we have $G = \{\pm e,\pm h_1,\ldots,\pm h_k\}$. On the other hand if G has odd order then $G \approx H$.

The double cover $S^3 \to SO(3)$ gives rise to finite subgroups of S^3 doubly covering those of $SO(3)$. Corresponding to D_{2n} we have D^*_{4n} of order $4n$

$$D^*_{4n} = \{x,y \mid x^2 = (xy)^2 = y^n\}$$

and corresponding to T, O, I we have the binary tetrahedral group T^* of order 24, the binary octahedral group O^* of order 48 and the binary icosahedral group I^* of order 120 presented by

$$\{x,y \mid x^2 = (xy)^3 = y^n\, , x^4 = 1\} \quad \text{for} \quad n = 3,4,5.$$

It can be shown that these are in fact the only finite subgroups of S^3 . Thus if $H_1 = e$ then G is one of these groups. Also, if H_1 is a cyclic group of relatively prime order to one of the above groups, then the direct product will act freely.

It remains to investigate the non-trivial possibilities. First note that if H is a subgroup of $H_1 \times H_2$ then the elements of the form $(h_1,e) \in H$ form a subgroup $H_1' \subset H_1$ and similarly $H_2' \subset H_2$ so that $H' = H_1' \times H_2' \subset H$ is an invariant subgroup. The quotient groups

$$H/H' \approx H_1/H_1' \approx H_2/H_2' \approx F$$

are isomorphic so H consists of elements (h_1,h_2) with the property that the coset of h_1 in H_1/H_1' corresponds to the coset of h_2 in H_2/H_2' under the isomorphism with F.

We again assume that $H_1 = C_m$ is cyclic.

If $H_2 = C_n$ is also cyclic, then we assert that H is also cyclic. This is clear if $(n,m) = 1$. Otherwise suppose that F is of order f so H_1' has order $m' = m/f$ and H_2' has order $n' = n/f$. Clearly they are also cyclic. We shall prove that if G acts freely on S^3, then H must also be cyclic. If a generates H_1 and b generates H_2 then H_1' consists of all powers of a^f and H_2' of b^f. Given an element of F, the elements of H_1 corresponding to it in the coset decomposition mod H_1' are those of the form $a^{kf+\rho}$ for fixed ρ and all possible k. If it corresponds to a generator of F, then its order is f and $(f,\rho) = 1$. Let k equal the product of all primes in m not in $f \cdot \rho$ (or $k = 1$ if no such prime exists). Then $(kf+\rho,m) = 1$ and $u = a^{kf+\rho}$ has order m and therefore generates H_1. We can find a similar generator v for H_2. It remains to show that (u,v) generates H. Since at least one of the preimages of (u,v) in $SO(4)$ is fixed point free, it follows from (6.1.2) that $(m',n') = 1$. Find p,q so that $pm' + qn' = 1$. Then clearly $pm \equiv f \pmod n$ and $qn \equiv f \pmod m$ so $u^{qn} = u^f$ and $v^{pm} = v^f$. From this we get for arbitrary k,l,ρ that

$$(u^{kf+\rho}, v^{lf+\rho}) = (u,v)^{kqn+lpm+\rho}$$

proving the assertion that H is cyclic.

Assuming that H_2 is one of the other groups D_{2m}, T,O,I and using similar arguments it can be shown that only two more types of groups occur.

If $H_1 = C_{2^k-1}$, $H_2 = D_{2(2n+1)}$, $H'_1 = C_{2^k-2}$, $H'_2 = C_{2n+1}$ and $H_1/H'_1 \approx H_2/H'_2 \approx C_2$ then we obtain a group H with double cover in $SO(4)$ equal to

$$D'_{2^k(2n+1)} = \{x,y \mid x^{2^k}=1,\ y^{2n+1}=1,\ xy^{-1}=y^{-1}x\} \quad k \geq 2,\ n \geq 1 .$$

Note that $D'_{4(2n+1)} = D^*_{4(2n+1)}$.

If $H_1 = C_{3^k}$, $H_2 = T$, $H'_1 = C_{3^{k-1}}$, $H'_2 = C_2 \times C_2$ and $H_1/H'_1 \approx H_2/H'_2 \approx C_3$ then we obtain a group H with double cover in $SO(4)$ equal to

$$T'_{8\cdot 3^k} = \{x,y,z \mid x^2 = (xy)^2 = y^2,\ zxz^{-1} = y,\ zyz^{-1} = xy,\ z^{3^k} = 1\}, k \geq 1.$$

Note that $T'_{24} = T^*_{24}$.

Thus we have the following conclusion, see H. Hopf [1], Seifert-Threlfall [1] and Milnor [2].

Theorem 1. The following is a list of all finite subgroups of $SO(4)$ that can act freely on S^3:

C_m, D^*_{4m}, $D'_{2^k(2n+1)}$, T^*, $T'_{8\cdot 3^k}$, O^*, I^* and the direct product of any of these groups with a cyclic group of relatively prime order.

Orbit spaces of finite groups acting freely and orthogonally on a sphere are called spherical Clifford-Klein manifolds. The 3-dimensional ones correspond to Seifert manifolds with finite fundamental group by (6.1.5) and are listed as follows, see Seifert-Threlfall [1].

Theorem 2. The Seifert manifolds with finite fundamental group are:

(i) $M = \{b;(o_1,0);(\alpha_1,\beta_1),(\alpha_2,\beta_2)\}$, here we allow $\alpha = 1$, $\beta = 0$, are lens spaces (see 5.4) with $\pi_1(M) = C_p$ where $p = |b\alpha_1\alpha_2 + \alpha_1\beta_2 + \beta_1\alpha_2|$;

(ii) $M = \{b;(o_1,0);(2,1),(2,1),(\alpha_3,\beta_3)\}$ are called prism manifolds. Let $m = (b+1)\alpha_3 + \beta_3$; if $(m,2\alpha_3) = 1$ then $\pi_1(M) = C_m \times D^*_{4\alpha_3}$, and if $m = 2m'$ then neccessarily m' is even and $(m',\alpha_3) = 1$ and letting $m' = 2^k m''$ we have $\pi_1(M) = C_{m''} \times D'_{2^{k+2}\alpha_3}$;

(iii) $M = \{b; (o_1,0);(2,1),(3,\beta_2),(3,\beta_3)\}$, let $m = 6b + 3 + 2(\beta_2+\beta_3)$, if $(m,12) = 1$ then $\pi_1(M) = C_m \times T^*$, and if $m = 3^k m'$, $(m',12) = 1$ then $\pi_1(M) = C_{m'} \times T'_{8\cdot 3^k}$;

(iv) $M = \{b;(o_1,0);(2,1),(3,\beta_2),(4,\beta_3)\}$, let $m = 12b + 6 + 4\beta_2 + 3\beta_3$, it follows that $(m,24) = 1$ and $\pi_1(M) = C_m \times O^*$;

(v) $M = \{b;(o_1,0);(2,1),(3,\beta_2),(5,\beta_3)\}$, let $m = 30b + 15 + 10\beta_2 + 6\beta_3$, it follows that $(m,60) = 1$ and $\pi_1(M) = C_m \times I^*$;

(vi) $M = \{b;(n_2,1);(\alpha_1,\beta_1)\}$ with $n = |b\alpha_1 + \beta_1| \neq 0$ are homeomorphic to prism manifolds so that
if α_1 is odd then $\pi_1(M) = C_{\alpha_1} \times D^*_{4n}$ and
if $\alpha_1 = 2^k\alpha_1'$, $(\alpha_1',2) = 1$ then $\pi_1(M) = C_{\alpha_1'} \times D'_{2^{k+2}n}$.

Proof. Except for (vi) the proof consists of verifying the group isomorphisms. It remains to prove that every prism manifold also admits a Seifert bundle structure of type n_2 over the projective plane. If G is the group, acting on S^3 with cyclic H_1 and $H_2 = D_{2n}$ the dihedral group then we consider the maximal cyclic subgroup C_n of D_{2n} and the cyclic group $C^*_{2n} \subset G$ mapping onto C_n . Since C^*_{2n} consists of left rotations ,

$C^*_{2n} \subset L \approx S^3$, it can be extended to a circle group $\Gamma \subset L$. If δ is a left rotation of order 4 in the group D^*_{4n} whose image is the reflection of D_{2n}, then for every element $\gamma \in \Gamma$ we have $\delta\gamma\delta^{-1} = \gamma^{-1}$. Thus δ maps the orbits of the circle action induced by Γ into each other reversing the orientation and S^3/G admits a Seifert fibration of class n_2. Since $\pi_1(M)$ is finite the orbit space is P^2 and $r \leq 1$.

Remark. It can be shown directly that apart from the lens spaces whose homeomorphism classification was given in (5.4) two 3-dimensional spherical Clifford-Klein manifolds are homeomorphic if and only if their fundamental groups are isomorphic. Note also that under (vi) $n = |b\alpha_1 + \beta_1| = 0$ if and only if $M = \{0;(n_2,1)\} = \mathbb{R}P^3 \# \mathbb{R}P^3$, see (5.4).

6.3. Non-orthogonal Actions

It is not known whether there exists a smooth free action of any group G on S^3 not conjugate to one of the orthogonal actions above. Since every such action has as orbit space a closed, orientable 3-manifold M with fundamental group G, it follows that G must have cohomology of period 4. We see from (6.1.2) that G can have at most one element of order 2. All finite groups not appearing in (6.2.1) satisfying these conditions are listed by Milnor [2] as follows:

(i) $Q(8n,k,l) = \{x,y,z \mid x^2 = (xy)^2 = y^{2n},\ z^{kl} = 1,\ xzx^{-1} = z^r, yzy^{-1} = z^{-1}\}$

where $8n,k,l$ are pairwise relatively prime integers so that if n is odd, then $n > k > l \geq 1$ and if n is even, then $n \geq 2$, $k > l \geq 1$.

(ii) $O'_{48 \cdot 3^k}$ $k \geq 1$ is the extension $1 \to C_{3^k} \to O'_{48 \cdot 3^k} \to O^* \to 1$ with the property that its 3-Sylow subgroup is cyclic and the action of O^* on C_{3^k} is given as follows: The commutator subgroup $T^* \subset O^*$ acts trivially, while the remaining elements of O^* carry each element of C_{3^k} into its inverse.

(iii) the product of any of these groups with a cyclic group of relatively prime order.

The smallest group on this list is $Q(16,3,1)$ of order 48 that may or may not be the fundamental group of a 3-manifold.

7. Fibering over S^1

In this chapter we shall find the Seifert manifolds that admit a locally trivial fibration with base S^1 and fiber a 2-manifold. This was originally done by Orlik-Vogt-Zieschang [1] for almost all cases and completed by Orlik-Raymond [2]. These results are recalled in section 2. In the meantime, however, a beautiful theory of injective toral actions has been developed by Conner-Raymond [1] and we shall discuss these general considerations first. Tollefson [1] and Jaco [1] noted independently that the product bundles $M = \{0;(o_1,g)\}$ fiber over S^1 in infinitely many distinct ways, i.e. with infinitely many mutually non-homeomorphic fibers. An outline of this argument is given in Section 3.

7.1. Injective Toral Actions

This section consist of results of Conner-Raymond [1].

Let X be paracompact, pathconnected, locally pathconnected and have the homotopy type of a CW complex. In the applications we shall assume that X is a manifold. An action of the torus group $T^k = S^1 \times S^1 \times \dots \times S^1$ (k times) on X is called <u>injective</u> if the map

$$f_*^x : \pi_1(T^k,1) \to \pi_1(X,x)$$

defined by $f_*^x(t) = tx$ is a monomorphism for all x.
In this case we have a central extension

$$0 \to \mathbb{Z}^k \to \pi_1(X) \to F \to 1$$

and only finite isotropy groups occur.

Theorem 1. *Let (T^k, X) be an action and $H_1(X;\mathbb{Z})$ be finitely generated. Then (T^k, X) fibers equivariantly over T^k if and only if the induced map*

$$f_*^x : H_1(T^k, 1) \to H_1(X, x)$$

is a monomorphism.

Note that if f_*^x is a monomorphism then so is $f_\#^x$ and the action is injective. For the proof we start with an injective action and consider subgroups of $\pi_1(X,x)$ containing $\mathrm{im}\, f_\#^x$. Let B_H be the covering space associated with H and $b_0 \in B_H$ be a base point corresponding to the constant path at x. The action of T^k may be lifted to B_H

$$\begin{array}{ccc} T^k \times B_H & \dashrightarrow & B_H \\ \downarrow & & \downarrow \\ T^k \times X & \longrightarrow & X \end{array}$$

since in the corresponding diagram of fundamental groups $\mathrm{im} f_\#^x \subset H$.

Theorem 2. *If $\mathrm{im}\, f_\#^x \subset H$ and H is normal then the action (T^k, B_H) is equivariantly homeomorphic to $(T^k, T^k \times Y)$, where the T^k action is just left translation on the first factor.*

The most important case is when $\varphi = \mathrm{id} : \pi_1(X,x) \to \pi_1(X,x)$ and $H = \mathrm{im}(f_\#^x)$. Note that in this case $\pi_1(B_H) = H = \mathbb{Z}^k$ so Y is simply connected.

The proof of theorem 2 consists of first showing that there is a natural splitting $H \simeq \mathbb{Z}^k \times \ker\varphi$. This follows because $h \in \pi_1(X,x)$ lies in H if and only if there is $t \in \mathbb{Z}^k$ so that $\varphi \circ f_\#^x(t) = \varphi(h) \in L$ and since $f_\#^x$ is a monomorphism t is unique. Define an epimorphism $p : H \to \mathbb{Z}^k$ by $p(h) = t$ in the above formula. We have $p(f^x_\#(t)) = t$ and $\ker\varphi = \ker p$. Define $q: H \to \ker\varphi$ by $q(h) = h\cdot f^x_\#(p(h^{-1}))$. Clearly $\operatorname{im} f^x_\# \subset \ker q$ and since it is a central subgroup it is the whole kernel. Note that if $h \in \ker\varphi$ then $q(h) = h$ and $h = f^x_\# p(h)\cdot q(h)$ proving the splitting of groups. Next we use induction on k. For $k = 1$ let ω be the generator of $\pi_1(S^1,1)$ represented by $\exp(2\pi i t)$, $0 \leq t \leq 1$. Then $f^{b_0}_\#(\omega) = \exp(2\pi i t)b_0$ represents the generator of the $\mathbb{Z}$ factor in $\pi_1(B_H) = H$ and by the naturality of the splitting b_0 must have trivial isotropy group, i.e. if $\exp(2\pi i t/n)b_0$, $0 \leq t \leq 1$, is a closed loop then necessarily $n = 1$. A similar argument applies for arbitrary $b \in B_H$ showing that the S^1-action is free. Induction on K proves that (T^k, B_H) is free. The fact that the principal T^k-bundle over B_H is trivial is obtained using the Leray-Hirsch theorem and the splitting $H \simeq \mathbb{Z}^k \times \ker\varphi$.

From the group of covering transformations $N = \pi_1(X,x)/H$ and the projection in the splitting onto Y we obtain an N-action on Y which turns out to be properly discontinuous (all isotropy groups are finite and the slice theorem holds).

The next step in the proof of theorem 1 is to classify actions of N on $T^k \times Y$ with the property that

(i) T^k acts on the first factor by left translations,

(ii) the action of N commutes with this T^k action and is equivariant with a given properly discontinuous action (N,Y) by the projection map.

Such actions are in one-to-one correspondence with elements of $H^1(N;\mathrm{Maps}(Y,T^k))$ where the N-module structure on the abelian group $\mathrm{Maps}(Y,T^k)$ is given by $(\alpha f)y = f(y\alpha)$ for $f \in \mathrm{Maps}(Y,T^k)$, $\alpha \in N$. Thus the action is given by a map $m: T^k \times Y \times N \to T^k$ so that for $t \in T^k$, $y \in Y$, $\alpha \in N$ we have $(t,y)\alpha = (m(t,y,\alpha),y\alpha)$. Now $m(t,y,\alpha) = tm(1,y,\alpha)$ by the left action of T^k so it is sufficient to consider maps $m: Y \times N \to T^k$ satisfying $m(y,\alpha\beta) = m(y,\alpha)m(y\alpha,\beta)$. The corresponding action is $(t,y)\alpha = (tm(y,\alpha),y\alpha)$. Consider these maps as $Z^1(N;\mathrm{Maps}(Y,T^k))$, the 1-dimensional cocycles. Two such maps $m_1(y,\alpha)$ and $m_2(y,\alpha)$ are cohomologous if they give rise to equivariant actions. Then there is a map $g: Y \to T^k$ so that we have an equivariant homeomorphism

$$F: (T^k, T^k \times Y, N)_1 \to (T^k, T^k \times Y, N)_2$$

defined by $F(t,y) = (tg(y),y)$ in which case

$$m_2(y,\alpha) = m_1(y,\alpha)g(y)g(y\alpha)^{-1} .$$

If the cohomology class of m is of finite order, say n, then there is a map $g: Y \to T^k$ for which

$$(*) \qquad g(y)g(y\alpha)^{-1} = m(y,\alpha)^n \qquad \text{for all } \alpha \in N .$$

In particular if N is a finite group of order n, then every element of $H^1(N;\mathrm{Maps}(Y,T^k))$ has finite order dividing n.

The last step in the proof of theorem 1 is to show that given the map g satisfying (*), the space X fibers over T^k with structure group $(\mathbb{Z}_n)^k$, where we think of $(\mathbb{Z}_n)^k \subset T^k$ as the product of n-th roots of unity. Let $C = \{(\tau,y) \mid \tau^n g(y) = 1\} \subset T^k \times Y$. It admits an action of $(\mathbb{Z}_n)^k$ since if $\lambda \in (\mathbb{Z}_n)^k$ and $(\tau,y) \in C$ then $(\lambda\tau,y) \in C$. Also, C is an invariant subset of the action $(T^k \times Y,N)$ because by (*) if $(\tau,y) \in C$ then $\tau^n m(y,\alpha)^n g(y\alpha) = \tau^n g(y) = 1$ showing that $(\tau m(y,\alpha),y\alpha) \in C$. Thus there are actions $((\mathbb{Z}_n)^k,C,N)$. Let $W = C/N$ with the induced $(\mathbb{Z}_n)^k$ action, let $[\tau,y] \in W$ be the equivalence class of (τ,y) under the action of N on C and $\pi: T^k \times Y \to X$ the N

orbit map. Define a new T^k-equivariant map $G: T^k \times W \to X$ by $G(t,[\tau,y]) = \pi(t\tau,y) = t\tau\pi(1,y)$. The fact that G is well defined follows from $\pi(t\tau m(y,\alpha),y\alpha) = t\tau\pi(1,y)$. If $G(t,[\tau,y]) = G(t_0,[\tau_0,y_0])$ then for some $\alpha \in N$ $y\alpha = y_0$ and $t\tau m(y,\alpha) = t_0\tau_0$. Now $t^n = t^n\tau^n m(y,\alpha)^n g(y\alpha)$ and $t_0^n = t_0^n\tau_0^n g(y_0) = t_0^n\tau_0^n g(y\alpha)$ so it follows that $t^n = t_0^n$ and therefore there is a $\lambda \in (\mathbb{Z}_n)^k$ such that $\lambda t_0 = t$, $\lambda\tau m(y,\alpha) = \tau_0$ and $(t\lambda^{-1},[\lambda\tau,y]) = (t_0,[\tau_0,y_0])$ showing that if $(\mathbb{Z}_n)^k$ acts on $T^k \times W$ by $\lambda(t,[\tau,y]) = (t\lambda^{-1},[\lambda\tau,y])$ then G induces a T^k-equivariant homeomorphism of $(T^k \times W)/(\mathbb{Z}_n)^k$ with X. The fibration over T^k is given by the map $(t,[\tau,y]) \to t^n$ with fiber W and structure group $(\mathbb{Z}_n)^k$.

The proof is completed by noting that if $f_*^x: H_1(T^k,1) \to H_1(X,x)$ is a monomorphism, then provided $H_1(X,x)$ is finitely generated, we have a direct summand L of rank k with $\operatorname{im} f_*^x \subset L$ and an epimorphism $\varphi: \pi_1(X,x) \to L$. The group $N = L/\varphi(\operatorname{im} f_{\#}^x)$ is therefore finite.

Observe that the construction depends on the choice of the map $g: Y \to T^k$. Different choices may even give fibers of different homotopy type as we shall show in section 3.

For X a closed 3-manifold and $k = 1$ we obtain the following statement.

<u>Corollary 3</u>. <u>A Seifert manifold</u> M <u>of class</u> o_1 <u>or</u> n_1 <u>admits an equivariant fibration over</u> S^1 <u>if and only if the order of the principal orbit</u> h <u>in</u> $H_1(M;\mathbb{Z})$ <u>is infinite</u>.

Note that if there is a fibration, then the characteristic map of the fiber (3.11) is of finite order. We shall see in the next section that large Seifert manifolds of the other classes do not admit a fibration over S^1, while some small Seifert manifolds admit non-equivariant fibrations over S^1 so that h has finite order in $H_1(M;\mathbb{Z})$ and the characteristic map is of infinite order.

7.2. Fibering Seifert Manifolds over S^1

A 3-manifolds is called irreducible if every tamely embedded 2-sphere bounds a 3-cell. The following result is due to Waldhausen [1], see (8.1).

Theorem 1. Large Seifert manifolds are irreducible.

The basic result on fibering 3-manifolds over S^1 is due to Stallings [1].

Theorem 2. Let M be an irreducible compact 3-manifold. If $\pi_1(M)$ has a finitely generated normal subgroup $N \neq \{1\},\mathbb{Z}_2$, with quotient $\pi_1(M)/N \approx \mathbb{Z}$ then M fibers over S^1 with fiber a compact 2-manifold T and $\pi_1(T) \approx N$.

These manifolds were classified by Neuwirth [1]. In particular for closed manifolds we have:

Theorem 3. Let M_2 be any closed irreducible 3-manifold and M_1 a closed manifold satisfying the conditions of theorem 2. Then M_1 is homeomorphic to M_2 if and only if $\pi_1(M_1)$ is isomorphic to $\pi_1(M_2)$.

The next result is from Orlik-Vogt-Zieschang [1].

Theorem 4. Let G be the fundamental group of a large Seifert manifold and H the maximal cyclic normal subgroup gene-

rated by h . *There is a finitely generated normal subgroup* $N \subset G$ *with* $G/N \approx \mathbb{Z}$ *if and only if* $[G,G] \cap H = \{1\}$.

Proof. If $[G,G] \cap H = \{1\}$ then H injects into $G/[G,G] = H_1(M;\mathbb{Z})$ and since it is an infinite cyclic subgroup of G its image is contained in an infinite summand of $G/[G,G]$. We can construct a homomorphism $\varphi\colon G \to \mathbb{Z}$ so that $\ker\varphi \cap H = \{1\}$. Then we have the commutative diagram

$$\begin{array}{ccccccccc} & & & & 0 & & & & \\ & & & & \downarrow & & & & \\ & & & & H & \searrow & & & \\ & & & & \downarrow{\scriptstyle i} & & & & \\ 1 & \longrightarrow & N & \longrightarrow & G & \xrightarrow{\varphi} & \mathbb{Z} & \longrightarrow & 0 \\ & & {\scriptstyle \psi}\downarrow & & \downarrow & & \downarrow & & \\ 1 & \longrightarrow & N' & \longrightarrow & G/H & \longrightarrow & \mathbb{Z}/\varphi H & \longrightarrow & 0 \end{array}$$

where N' is the kernel of the induced map $G/H \to \mathbb{Z}/\varphi H$. Since $\ker\varphi \cap H = \{1\}$ we see that ψ is an isomorphism. But G/H is finitely generated and $\mathbb{Z}/\varphi H$ is finite so N' and hence N is finitely generated. Note that this argument has elements of the proof of (7.1.1).

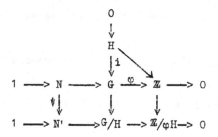

Conversely, if N is a finitely generated normal subgroup with $G/N \approx \mathbb{Z}$ then it follows from the fact that M is large and from the above theorem of Stallings that N is the fundamental group of a closed 2-manifold. If $N \cap H \neq \{1\}$ then N contains an infinite cyclic normal subgroup. This is only possible for the torus and the Klein bottle. Let $N' = N$ for the torus and let N' be the free abelian subgroup of rank 2 in N for the Klein bottle. Clearly, $N' \cap H \neq 1$ and $N'/N' \cap H$ must be a cyclic group since in G/H (M large!) two elements commute if and only if they are the powers of some other element. On the other hand $N'/N' \cap H$ would be a cyclic normal subgroup of G/H and this is a contradiction. Thus $N \cap H = \{1\}$ and clearly $[G,G] \cap H = \{1\}$.

Corollary 5. Let M be a large Seifert manifold. It fibers over S^1 if and only if the order of the fiber h in $H_1(M;\mathbb{Z})$ is infinite.

Since for classes other than o_1 and n_1 we have the homology relation $2h = 0$, this corollary gives the same condition as (7.1.3).

Looking at the homology relations one can see immediately (3.11) that

(i) for o_1 the order of h is infinite in $H_1(M;\mathbb{Z})$ if and only if

$$p = b\alpha_1 \ldots \alpha_r + \beta_1\alpha_2 \ldots \alpha_r + \ldots + \alpha_1 \ldots \alpha_{r-1}\beta_r = 0$$

(ii) for n_1 the order of h is always infinite in $H_1(M;\mathbb{Z})$.

For a manifold M let $\Lambda(M)$ denote its homeotopy group, the group of isotopy classes of self-homeomorphisms divided by the subgroup of those isotopic to the identity. For a group G we denote by $\mathrm{Aut}(G)$ the full group of automorphisms of G and by $\mathrm{In}(G)$ the subgroup of inner automorphisms.

If M is a B-bundle over S^1, then it is determined by the characteristic map $\Phi : B \to B$. If $B \neq S^2, P^2$ then theorem 3 says that M is determined by its fundamental group. Now a well-known theorem of Nielsen states that

$$\Lambda(B) = \mathrm{Aut}(\pi_1 B)/\mathrm{In}(\pi_1 B)$$

so the isotopy class of Φ is determined by the induced automorphism $\varphi: \pi_1(B) \to \pi_1(B)$ up to inner automorphisms.

Given an automorphism of $\pi_1(B)$ we call the manifold, obtained as a fiber bundle over S^1 with characteristic map some Φ whose induced map agrees with φ up to an inner automorphism, M_φ. From the previous discussion it follows that M_φ is well defined. We let

$$\pi_1(B) = (x_1,\ldots,x_n|\pi_*)$$

where $\pi_* = [x_1,x_2],\ldots,[x_{n-1},x_n]$ if B is orientable and $\pi_* = x_1^2,\ldots,x_n^2$ if B is non-orientable. A presentation of $\pi_1(M_\varphi)$ is then given by

$$\pi_1(M_\varphi) = (x_1,\ldots,x_n,x|\pi_*,xx_ix^{-1} = \varphi(x_i),\ i = 1,\ldots,n).$$

Now consider the small Seifert manifolds, see Orlik-Raymond [2]. The two fibers we shall encounter are the torus T and the Klein-bottle, K. Recall that $\Lambda(T)$ is isomorphic to the multiplicative group of unimodular 2×2 integer entry matrices. It can be generated by

$$\varphi_1 = \begin{pmatrix} 0 & -1 \\ 1 & 0 \end{pmatrix},\ \varphi_2 = \begin{pmatrix} 0 & -1 \\ 1 & 1 \end{pmatrix},\ \varphi_3 = \begin{pmatrix} 0 & 1 \\ 1 & 0 \end{pmatrix}$$

and a presentation is given by

$$\Lambda(T) = (\varphi_1,\varphi_2,\varphi_3|\varphi_1^4 = \varphi_2^6 = \varphi_3^2 = \varphi_1^2\varphi_2^3 = (\varphi_1\varphi_3)^2 = (\varphi_2\varphi_3)^2 = 1).$$

The orientation preserving automorphisms (matrices with determinant +1) form a subgroup of index 2

$$\Lambda^+(T) = (\varphi_1,\varphi_2|\varphi_1^4 = \varphi_2^6 = \varphi_1^2\varphi_2^3 = 1)$$

isomorphic to the free product of C_4 and C_6 amalgamated along the subgroups isomorphic to C_2. This shows that the only elements of finite order in $\Lambda^+(T)$ are powers of φ_1 and φ_2 and their conjugates.

It is known that $\Lambda(K) = \mathbb{Z}_2 + \mathbb{Z}_2$ and generators may be given as the following automorphisms of $\pi_1(K) = (x_1, x_2 | x_1^2 x_2^2 = 1)$:

$$\psi_1(x_1) = x_2 \ , \ \psi_1(x_2) = x_1 \ ; \ \psi_2(x_1) = x_1^{-1} \ , \ \psi_2(x_2) = x_2^{-1} \ .$$

Now let us consider the small Seifert manifolds.

(i) $o_1, g = 0$, $r \leq 2$ are either lens spaces or $S^2 \times S^1$, the latter if and only if $p = b\alpha_1\alpha_2 + \beta_1\alpha_2 + \alpha_1\beta_2 = 0$. From this equation we conclude that $\alpha_2 = \alpha_1$ and $\beta_2 = -(b\alpha_1 + \beta_1)$ so $b = -1$ and $\beta_2 = \alpha_1 - \beta_1$. Thus the complete set of S^1-actions on $S^2 \times S^1$ is given by the collection $\{-1;(o_1,0),(\alpha_1,\beta_1),(\alpha_1,\alpha_1-\beta_1)\}$. The order of h is infinite in $H_1(S^2 \times S^1;\mathbb{Z})$.

(ii) $o_1, g = 0$, $r = 3$, $\frac{1}{\alpha_1} + \frac{1}{\alpha_2} + \frac{1}{\alpha_3} > 1$ have finite $H_1(M;\mathbb{Z})$ and cannot fiber over S^1 .

(iii) $M = \{-2;(o_1,0);(2,1),(2,1),(2,1),(2,1)\}$ satisfies the condition for an injective action and it is easily seen that h has infinite order in $H_1(M;\mathbb{Z})$. In fact there is an equivariant fibration of M over S^1 with fiber T and $\varphi = \varphi_1^2 \in \Lambda^+(T)$, see (ix) below.

(iv) $M = \{b;(o_1,1)\}$ are T-bundles over S^1 . Specifically, $\pi_1(M) = (a_1, b_1, h | [a_1,b_1]h^{-b}, [a_1,h], [b_1,h])$ and the map $f(a_1) = x_1$, $f(b_1) = x$, $f(h) = x_2$ defines an isomorphism with M_φ for $\varphi = (\varphi_1^3\varphi_2)^{-b} \in \Lambda^+(T)$ whose matrix is $\begin{pmatrix} 1 & -b \\ 0 & 1 \end{pmatrix}$. Note in particular that for $b \neq 0$ φ has infinite order in $\Lambda^+(T)$ and h has finite order in $H_1(M;\mathbb{Z})$. Of course, for $b = 0$ we have $M = S^1 \times S^1 \times S^1$.

(v) $M = \{b;(o_2,1)\}$ are two of the four K-bundles over S^1.

With the notation above we have

$$\{0;(o_2,1)\} = M_{id} = K \times S^1 \quad \text{and} \quad \{1;(o_2,1)\} = M_{\psi_1\psi_2}$$

by $f(a_1) = x_1$, $f(b_1) = x_1^{-1}x$, $f(h) = x_1x_2$.

(vi) n_1, $g = 1$, $r \leq 1$ give the possible S^1 actions on $P^2 \times S^1$ and N and both fiber over S^1 .

(vii) n_2, $g = 1$, $r \leq 1$ are the prism manifolds with finite fundamental groups and $\{0;(n_2,1)\} = \mathbb{R}P^3 \# \mathbb{R}P^3$ so they do not fiber over S^1 .

(viii) $M = \{b;(n_1,2)\}$ are the same two K-bundles over S^1 as under (v),

$$\{0;(n_1,2)\} = M_{id} = K \times S^1 \quad \text{and} \quad \{1;(n_1,2)\} = M_{\psi_1\psi_2} .$$

The first is obvious. The second is given by $f(v_1) = x_1$, $f(v_2)=x$, $f(h) = x_2^{-2}x^2$.

(ix) $M = \{b;(n_2,2)\}$ are T-bundles over S^1 . Specifically, $\pi_1(M) = (v_1,v_2,h \mid v_1^2v_2^2h^{-b},\ v_1hv_1^{-1}h,\ v_2hv_2^{-1}h)$ and the map $f(v_1) = x$, $f(v_2)=x_1x^{-1}$, $f(h) = x_2$ defines an isomorphism with M_φ for $\varphi = \varphi_1^2(\varphi_1^3\varphi_2)^b \in \Lambda^+(T)$ whose matrix is $\begin{pmatrix} -1 & -b \\ 0 & -1 \end{pmatrix}$. For $b \neq 0$ the order of φ is infinite and $\pi_1(M)$ is centerless. For $b = 0$ the manifold $\{0;(n_2,2)\}$ is homeomorphic to $\{-2;(o_1,0);(2,1),(2,1),(2,1),(2,1)\}$ as noted in (5.4). Thus the latter is also a T-bundle over S^1 with characteristic map of order 2 and matrix $\begin{pmatrix} -1 & 0 \\ 0 & -1 \end{pmatrix}$.

(x) $M = \{b;(n_3,2)\}$ are the other two K-bundles over S^1 ,

$$\{0;(n_3,2)\} = M_{\psi_2} \quad \text{and} \quad \{1;(n_3,2)\} = M_{\psi_1} .$$

The first is given by $f(v_1) = x_1x^{-1}$, $f(v_2) = x$, $f(h) = x_1^{-1}x_2^{-1}$, the second by $f(v_1) = x$, $f(v_2) = x^{-1}x_1$, $f(h) = x_2x_1$.

7.3. Non-uniqueness of the Fiber

The choice of the map $g: Y \to T^k$ in the proof of (7.1.1) determines the fiber. The non-uniqueness is clearly seen by the following example of Tollefson [1].

Let $T(m)$ denote a closed orientable 2-manifold of genus $m = k(g-1)+1$ where $g > 1$ and arrange $T(m)$ in $\mathbb{R}^3$ with k arms each of genus $(g-1)$ about one hole at the origin, see picture below for $k = 3$, $g = 3$.

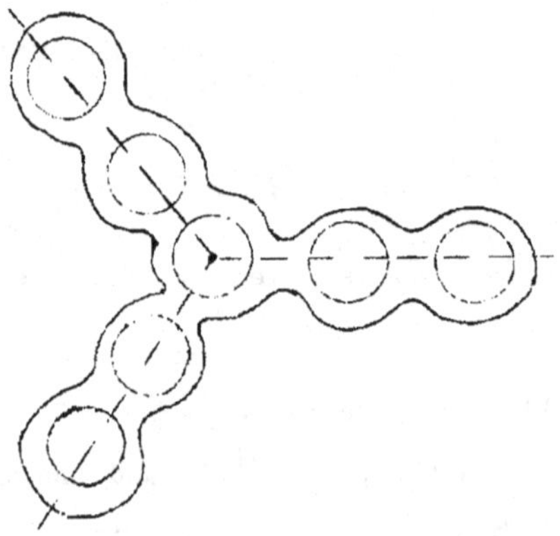

Let $\varphi: T(m) \to T(m)$ generate a free $\mathbb{Z}_k$ action by rotating through the angle $2\pi i/k$ and consider the 3-manifold M that is a $T(m)$-bundle over S^1 with characteristic map φ. It admits an obvious free S^1-action as follows: If $[x,t] \in T(m)\times I/(x,0) = (\varphi(x),1)$ is the equivalence class of a point and $s \in S^1 = \mathbb{R}/\mathbb{Z}$ then define

$$[s]([x,t]) = [x,t+ks] .$$

The action is equivariant with respect to the $\mathbb{Z}_k$ action and its orbit space is $T(g)$. Thus $M = \{b;(o_1,g)\}$ and since it fibers over S^1, it follows from (7.2.4) that $b = 0$, hence $M = T(g) \times S^1$.

Thus for $m = k(g-1)+1$ we can embed $T(m)$ in $T(g) \times S^1$ as a non-separating surface with complement $T(m) \times I$ so that the projection map $p: T(g) \times S^1 \to T(g)$ restricted to $T(m)$ is a covering. A much stronger statement about incompressible surfaces in S^1-bundles due to Waldhausen [1] may be found in (8.1.3).

8. Further Topics

The important results of Waldhausen [1,2,3] occupy a central position in the theory of 3-manifolds in general and Seifert manifolds in particular. It would carry us too far afield to give a detailed account of his work so we have to restrict ourselves in section 1 to a description of the most relevant results. In his book Wolf [1] determines all closed 3-dimensional flat riemannian manifolds. There are six orientable and four non-orientable such manifolds and in section 2 we identify them as Seifert manifolds. Section 3 lists Seifert manifolds with solvable fundamental groups as determined by L. Moser [1]. We consider finite groups acting on Seifert manifolds in section 4. Some remarks on foliations in section 5 and on flows in section 6 conclude the notes.

8.1. Waldhausen's Results

Waldhausen [1,2,3] works in the piecewise linear category so manifolds have combinatorial triangulations, submanifolds are subcomplexes and maps are piecewise linear. Manifolds are always orientable compact 3-manifolds and may have boundaries. Regular neighborhoods of submanifolds are also compact and chosen sufficiently small with respect to the already given submanifolds of the manifold in question. In general the embedding of a surface F in a manifold M is proper, $F \cap \partial M = \partial F$ and F is orientable, hence 2-sided. A system of surfaces has a finite number of disjoint components. Homeomorphisms are assumed to be surjective. An isotopy of X is a level preserving map $h: X \times I \to X \times I$ so that at each level $h|X \times t = h_t\colon X \to X$ is a homeomorphism. We shall assume that $h_0 = \mathrm{id}$ and call an isotopic deformation simply a <u>deformation</u>. Two subspaces of X, Y_1 and Y_2 are <u>isotopic</u> if there is an ambient isotopy of X so that $h_1(Y_1) = Y_2$. Two surfaces F and G in M or ∂M with $F \cap G = \partial F = \partial G$ are called <u>parallel</u> if there is a surface H and embedding $f\colon H \times I \to M$ so that $f(H \times 0) = F$ and $f(H \times 1 \cup \partial H \times I) = G$. A surface F in M is called ∂-<u>parallel</u> (boundary-parallel) if there is a surface F in ∂M parallel to F. For curves in sufaces we define parallel and ∂-parallel similarly.

The following construction is often repeated. Given a system of surfaces F in M a new (not necessarily connected) manifold $\tilde{M}$ is obtained by <u>cutting up</u> M <u>along</u> F, i.e. let $U(F)$ be a regular neighborhood of F in M and let $\tilde{M} = \overline{M - U(F)}$. We can thus view $\tilde{M}$ as a submanifold of M. Note that the construction is well defined up to an isotopy of F. Given another system of surfaces G in M in general position w.r.t. F, the new system $\tilde{G} = G \cap \tilde{M}$, however, depends on prior deformations of F.

A system of surfaces F in M or ∂M is <u>compressible</u> if one of the following holds:

(i) there is a simple closed curve k in $\mathring{F}$ that does not bound a 2-cell in F and an embedding of a 2-cell D in M so that $\mathring{D} \subset \mathring{M}$ and $D \cap F = k$;

(ii) there is an embedding of a 3-cell E in M so that $E \cap F = \partial E$.

The negation of compressible is denoted <u>incompressible</u>. Thus M is <u>irreducible</u> if it contains no <u>incompressible</u> 2-sphere. Here are some of the main results of Waldhausen [1]:

Theorem 1. *Let* F *be an incompressible system of surfaces in* M *and* $\tilde{M} = \overline{M - U(F)}$. $\tilde{M}$ *is irreducible if and only if* M *is irreducible.*

Let B be a compact, not necessarily orientable 2-manifold and $p: M \to B$ an S^1-bundle over B with orientable total space. Thus if M is closed it is a Seifert manifold of class o_1 or n_2 . A subspace $X \subset M$ is *vertical* if $X = p^{-1}(p(X))$ and *horisontal* if $p|X$ is an embedding.

Lemma 2. *Let* $p: M \to B$ *be an* S^1*-bundle. If* B *is not* S^2 *or* P^2 *then* M *is irreducible.*

Note that the S^1-bundles over S^2 are lens spaces and known to be irreducible or $S^2 \times S^1$ while the S^1-bundles over P^2 are prism manifolds and irreducible or $\{0;(n_2,1)\} = \mathbb{R}P^3 \# \mathbb{R}P^3$. If a manifold has irreducible orientable double cover, then it is itself irreducible so the above lemma proves the irreducibility of all S^1-bundles with the noted exceptions, $P^2 \times S^1$ and N .

Theorem 3. *Let* $p: M \to B$ *be an* S^1*-bundle where* B *is not* S^2 *or* P^2 . *Let* G *be a system of incompressible surfaces in* M *so that no bounded component of* G *is* ∂*-parallel. Then there is an ambient isotopy so that the result is either that*

(i) G *is vertical so each component of* G *is an annulus or a torus; or*

(ii) $p|G$ *is a covering map.*

The basic result on the homeomorphisms of S^1-bundles is the following:

Theorem 4. *Let* $p: M \to B$ *and* $p': M' \to B'$ *be* S^1*-bundles.*

<u>Assume that neither</u> B <u>nor</u> B' <u>is</u> S^2, P^2, D^2 <u>or</u> $S^1 \times I$ <u>and if</u> B <u>or</u> B' <u>is the torus or Klein bottle then the bundle has no cross-section</u>. <u>Let</u> $\varphi: M \to M'$ <u>be a homeomorphism</u>. <u>There exists a homeomorphism</u> $\psi: M \to M'$ <u>so that</u>

(i) ψ <u>is isotopic to</u> φ ,

(ii) <u>there is a map</u> $p(\psi): B \to B'$ <u>making</u> $(\psi, p(\psi))$ <u>a bundle isomorphism</u>.

Given a manifold M, a system of tori $T = T_1 \cup \ldots \cup T_n$, $n \geq 0$ in the interior of M with regular neighborhood U(T) is called a <u>graph structure</u> ("Graphenstruktur") on M if $M - \mathrm{int}\, U(T)$ is an S^1-bundle. M is then called a graph manifold ("Graphenmannigfaltigkeit"). In order to define a simple graph structure let T_1 be a component of T and $U(T_1)$ its regular neighborhood homeomorphic to torus x interval with boundary components T' and T". Let M_1 be the component of $M - \mathrm{int}\, U(T)$ meeting T' and M_2 meeting T". The natural isomorphisms

$$H_1(T') \longleftrightarrow H_1(U(T_1)) \longleftrightarrow H_1(T'')$$

allow us to talk about intersections of homology classes of curves on T' and T". A graph structure is <u>simple</u> (and the graph manifold is simple) if it is not one of the following:

(i) M_1 is not identical to M_2 and M_1 is the bundle over the annulus,

(ii) the fiber of M_1 is homologous to the fiber of M_2 ,

(iii) M_1 is a solid torus and a meridian curve has intersection number 1 with a fiber of M_2 ,

(iv) M_1 is a solid torus and a meridian curve is homologous to a fiber of M_2 ,

(v) M_1 is the S^1-bundle over the Moebius band and we

think of it embedded as a cross-section in M_1 so that its boundary is homologous to the fiber in M_2 ,

(vi) both M_1 and M_2 are S^1-bundles over the Moebius band with embedded cross-sections whose boundaries are homologous,

(vii) $M - \operatorname{int} U(T_1)$ has two components, one called Q is obtained by sewing two orbits of type (2,1) into $D^2 \times S^1$ and the other is not a solid torus,

(viii) M_1 and M_2 are identical and isomorphic to torus x interval and the composition of natural isomorphisms

$$H_1(T') \to H_1(U(T_1)) \to H_1(T'') \to H_1(M_1) \to H_1(T')$$

maps an element onto itself or its inverse,

(ix) M_1 and M_2 are solid tori,

(x) $T = \emptyset$ and M is a bundle over S^2 or P^2 .

Waldhausen [1] gives a complete classification of graph manifolds up to homeomorphism and shows that Seifert manifolds are special cases of graph manifolds. Here are the main results.

Theorem 5. A simple graph manifold is irreducible.

Theorem 6. Let M and N be simple graph manifolds with graph structures $T = T_1 \cup \ldots \cup T_m$ and $T' = T_1' \cup \ldots \cup T_n'$. Assume that the pair (M,N) is not one of the exceptions below. Then given a homeomorphism $\varphi: M \to N$ there exists an isotopic homeomorphism $\psi: M \to N$ so that $\psi(T) = T'$.

Exceptions:

(i) $\tilde{M} = M - \operatorname{int} U(T)$ is a bundle over the m-holed 2-sphere and m solid tori with $m \leq 3$; or $\tilde{M}$ is a bundle over the m-holed projective plane and m solid tori with $m \leq 1$. The same for $\tilde{N} = N - \operatorname{int} U(T')$.

(ii) $\tilde{M} = M - \text{int}\, U(T)$ is torus x interval and $\tilde{N} = N - \text{int}U(T')$ is a bundle over the n-holed 2-sphere and n solid tori with $n \leq 3$ - or vica versa.

(iii) M is the manifold Q above and N is the S^1-bundle over the Moebius band - or vica versa.

(iv) $M = \{-2;(o_1,0);(2,1),(2,1),(2,1),(2,1)\}$, $N = \{0;(n_2,2)\}$ - or vica versa.

We shall call an orientable Seifert manifold <u>sufficiently large</u> if it is not on the list below.

(i) o_1, $g = 0$, $r \leq 2$

(ii) o_1, $g = 0$, $r = 3$

(iii) n_2, $g = 1$, $r \leq 1$

(iv) $S^1 \times S^1 \times S^1$

(v) $\{0;(n_2,2)\}$

(vi) $\{-2;(o_1,0);(2,1),(2,1),(2,1),(2,1)\}$

(vii) $\{-1;(n_2,1);(2,1),(2,1)\}$

A corollary of theorem 6 is the following result.

<u>Theorem 7</u>. <u>Let</u> M <u>and</u> N <u>be sufficiently large orientable Seifert manifolds</u>. <u>Given a homeomorphism</u> $\varphi: M \to N$ <u>there exists an isotopic homeomorphism</u> $\psi: M \to N$ <u>so that</u> ψ <u>induces a Seifert bundle isomorphism</u>.

The proof consists of showing that if we take a simple closed curve about each component of E^* in M^* (and N^*) and consider their inverse images, then this collection of tori gives rise to a simple graph structure on M (and N). In particular this proves the irreducibility of these manifolds up to a few exceptions as claimed in (7.2.1).

This is considerably stronger than (5.3.6) where we showed only the existence of <u>some</u> Seifert bundle isomorphism. Much more is true, however. According to Waldhausen [2] two irreducible, sufficiently large closed orientable 3-manifolds are homeomorphic if their fundamental groups are isomorphic. The notion of "sufficiently large" means that M is not a ball and contains an incompressible surface. Equivalently, an irreducible closed manifold M is sufficiently large if and only if $H_1(M)$ is infinite or $\pi_1(M)$ is a non-trivial free product with amalgamation. For orientable Seifert manifolds the notion coincides with the definition above. As a corollary to this result of Waldhausen [2] we may state:

Theorem 8. Let M be a sufficiently large orientable Seifert manifold and N an irreducible, closed, orientable 3-manifold. If there exists an isomorphism $\varphi: \pi_1 M \to \pi_1 N$ then there exists a homeomorphism $\Phi: M \to N$ inducing φ.

Waldhausen [2] also makes some comments about the homeotopy group $\Lambda(M)$ of M. The following Nielsen-type theorem holds for sufficiently large manifolds but will be stated here only for Seifert manifolds.

Theorem 9. Let M be a sufficiently large Seifert manifold. Then there is a natural isomorphism

$$\Lambda(M) \approx \mathrm{Aut}(\pi_1 M)/\mathrm{In}(\pi_1 M) .$$

Letting $\Gamma(M)$ denote the group of fiber preserving homeomorphisms of M modulo those that are isotopic to the identity by fiber preserving isotopies, Waldhausen [2] shows that the natural map

$$\Gamma(M) \to \Lambda(M)$$

is an isomorphism for sufficiently large Seifert manifolds.

Surjectivity follows from theorem 7 and injectivity from the methods developed in Waldhausen [2]. It requires deforming an isotopy into a fiber preserving isotopy. Not much is known about the structure of $\Gamma(M)$.

Recall that if the orientable Seifert manifold M admits an S^1-action, then h is in the center of $\pi_1(M)$. The following remarkable conversion of this fact is obtained in Waldhausen [3].

Theorem 10. Let M be an irreducible, closed, orientable, sufficiently large 3-manifold. If $\pi_1(M)$ has a non-trivial center then M is homeomorphic to a Seifert manifold of class o_1 and therefore admits an S^1-action.

Several of these results may be extended to non-orientable Seifert manifolds by lifting to the orientable double cover. Let $M = \{b;(\epsilon,g);(\alpha_1,\beta_1),\ldots,(\alpha_r,\beta_r)\}$ be a non-orientable Seifert manifold. According to Seifert [1] its orientable double cover is

$$\tilde{M} = \{-r;(\hat{\epsilon},\hat{g});(\alpha_1,\beta_1),\ldots,(\alpha_r,\beta_r),(\alpha_1,\alpha_1-\beta_1),\ldots,(\alpha_r,\alpha_r-\beta_r)\}$$

where

ϵ	o_2	n_1	n_3	n_4
$\hat{\epsilon}$	o_1	o_1	n_2	n_2
$\hat{g}$	2g-1	g-1	2g-2	2g-2

.

8.2. Flat Riemannian Manifolds

In this section we shall identity as Seifert manifolds the closed flat riemannian 3-manifolds found by Wolf [1]. Let E(n) denote the group of rigid motions of R^n . Every rigid motion consists of a translation, t_a by a vector a followed by a rotation A . Write the motion (A, t_a) . Clearly A is an element of $O(n)$ and a is an arbitrary vector in R^n . Thus the <u>euclidean group</u> $E(n)$ is the semi-direct product of $O(n)$ and R^n satisfying the following product rule:

$$(A, t_a)(B, t_b) = (AB, t_{Ab+a}) .$$

We write $E(n) = O(n) \cdot R^n$. Obviously $E(n)$ is a Lie group acting on R^n and $R^n = E(n)/O(n)$ as coset space.

<u>A flat compact, connected riemannian manifold</u> M^n is the orbit space of R^n by the free properly discontinuous action of a discrete subgroup $\Gamma \subset E(n)$, $M^n = R^n/\Gamma$. It admits a covering by the torus T^n . The group Γ has an abelian normal subgroup Γ^* of rank n and finite index. As a group $\Gamma^* = \Gamma \cap R^n$. It follows also that Γ has no non-trivial element of finite order. The group of deck transformations Ψ in the covering $T^n \to M^n$ is called the <u>holonomy group</u> of M^n , $\Psi = \Gamma/\Gamma^*$.

The following result is from Wolf [1,p.117].

<u>Theorem 1. There are just</u> 6 <u>affine diffeomorphism classes of compact connected orientable flat</u> 3-<u>dimensional riemannian manifolds. They are represented by the manifolds</u> R^3/Γ <u>where</u> Γ <u>is one of the six groups</u> $\mathcal{G}_i$ <u>given below. Here</u> Λ <u>is the translation lattice,</u> $\{a_1, a_2, a_3\}$ <u>are its generators,</u> $t_i = t_{a_i}$, <u>and</u> $\Psi = \Gamma/\Gamma^*$ <u>is the holonomy</u>.

$\mathcal{G}_1$. $\Psi = \{1\}$ <u>and</u> Γ <u>is generated by the translations</u> $\{t_1, t_2, t_3\}$ <u>with</u> $\{a_i\}$ <u>linearly independent</u>.

$\mathcal{G}_2$. $\Psi = \mathbb{Z}_2$ <u>and</u> Γ <u>is generated by</u> $\{\alpha, t_1, t_2, t_3\}$ <u>where</u> $\alpha^2 = t_1$, $\alpha t_2 \alpha^{-1} = t_2^{-1}$ <u>and</u> $\alpha t_3 \alpha^{-1} = t_3^{-1}$; a_1 <u>is orthogonal to</u> a_2 <u>and</u> a_3 <u>while</u> $\alpha = (A, t_{a_1/2})$ <u>with</u> $A(a_1) = a_1$, $A(a_2) = -a_2$, $A(a_3) = -a_3$.

$\mathcal{G}_3$. $\Psi = \mathbb{Z}_3$ and Γ is generated by $\{\alpha, t_1, t_2, t_3\}$ where $\alpha^3 = t_1$, $\alpha t_2 \alpha^{-1} = t_3$ and $\alpha t_3 \alpha^{-1} = t_2^{-1} t_3^{-1}$; a_1 is orthogonal to a_2 and a_3, $\|a_2\| = \|a_3\|$ and $\{a_2, a_3\}$ is a hexagonal plane lattice, and $\alpha = (A, t_{a_1/3})$ with $A(a_1) = a_1$, $A(a_2) = a_3$, $A(a_3) = -a_2 - a_3$.

$\mathcal{G}_4$. $\Psi = \mathbb{Z}_4$ and Γ is generated by $\{\alpha, t_1, t_2, t_3\}$ where $\alpha^4 = t_1$, $\alpha t_2 \alpha^{-1} = t_3$ and $\alpha t_3 \alpha^{-1} = t_2^{-1}$; $\{a_i\}$ are mutually orthogonal with $\|a_2\| = \|a_3\|$ while $\alpha = (A, t_{a_1/4})$ with $A(a_1) = a_1$, $A(a_2) = a_3$, $A(a_3) = -a_2$.

$\mathcal{G}_5$. $\Psi = \mathbb{Z}_6$ and Γ is generated by $\{\alpha, t_1, t_2, t_3\}$ where $\alpha^6 = t_1$, $\alpha t_2 \alpha^{-1} = t_3$, $\alpha t_3 \alpha^{-1} = t_2^{-1} t_3$; a_1 is orthogonal to a_2 and a_3, $\|a_2\| = \|a_3\|$ and $\{a_2, a_3\}$ is a hexagonal plane lattice, and $\alpha = (A, t_{a_1/6})$ with $A(a_1) = a_1$, $A(a_2) = a_3$, $A(a_3) = a_3 - a_2$.

$\mathcal{G}_6$. $\Psi = \mathbb{Z}_2 \times \mathbb{Z}_2$ and Γ is generated by $\{\alpha, \beta, \gamma; t_1, t_2, t_3\}$ where $\gamma\beta\alpha = t_1 t_3$ and

$$\alpha^2 = t_1, \quad \alpha t_2 \alpha^{-1} = t_2^{-1}, \quad \alpha t_3 \alpha^{-1} = t_3^{-1}$$
$$\beta t_1 \beta^{-1} = t_1^{-1}, \quad \beta^2 = t_2, \quad \beta t_3 \beta^{-1} = t_3^{-1}$$
$$\gamma t_1 \gamma^{-1} = t_1^{-1}, \quad \gamma t_2 \gamma^{-1} = t_2^{-1}, \quad \gamma^2 = t_3$$

The $\{a_i\}$ are mutually orthogonal and

$\alpha = (A, t_{a_1/2})$ with $A(a_1) = a_1$, $A(a_2) = -a_2$, $A(a_3) = -a_3$;

$\beta = (B, t_{(a_2+a_3)/2})$ with $B(a_1) = -a_1$, $B(a_2) = a_2$, $B(a_3) = -a_3$;

$\gamma = (C, t_{(a_1+a_2+a_3)/2})$ with $C(a_1) = -a_1$, $C(a_2) = -a_2$, $C(a_3) = a_3$.

Theorem 2. The six compact, connected orientable flat riemannian 3-manifolds of theorem 1 are the Seifert manifolds:

$M_1 = \{0;(o_1,1)\} = S^1 \times S^1 \times S^1$;

$M_2 = \{-2;(o_1,0);(2,1),(2,1),(2,1),(2,1)\}$ is the T^2 bundle over S^1 with matrix of the characteristic map $\begin{pmatrix} -1 & 0 \\ 0 & -1 \end{pmatrix}$ of order 2;

$M_3 = \{-1;(o_1,0);(3,1),(3,1),(3,1)\}$ is the T^2 buntle over S^1 with matrix of the characteristic map $\begin{pmatrix} 0 & 1 \\ -1 & -1 \end{pmatrix}$ of order 3 ;

$M_4 = \{-1;(o_1,0);(2,1),(4,1),(4,1)\}$ is the T^2 bundle over S^1 with matrix of the characteristic map $\begin{pmatrix} 0 & 1 \\ -1 & 0 \end{pmatrix}$ of order 4 ;

$M_5 = \{-1;(o_1,0);(2,1),(3,1),(6,1)\}$ is the T^2 bundle over S^1 with matrix of the characteristic map $\begin{pmatrix} 0 & 1 \\ -1 & 1 \end{pmatrix}$ of order 6 ;

$M_6 = \{-1;(n_2,1);(2,1),(2,1)\}$ is the manifold obtained from taking the two Seifert fibrations of Q , one as a solid torus with two orbits of type $(2,1)$ and the other as the circle bundle over the Moebius band with orientable total space, and sewing them together by a fiber preserving homeomorphism. It is also the orbit space of the orientation preserving free involution on the Seifert bundle over S^2 with total space M_2 which identifies fibers over antipodal points of the base space by an orientation reversing homeomorphism.

Proof. Let $G_i = \pi_1(M_i)$. It suffices to show that $\mathcal{G}_i \cong G_i$ for $i = 1,\dots,6$. It will be clear from the isomorphisms in the first five cases that there is an S^1 action on $S^1 \times S^1 \times S^1$ making the action of the holonomy group equivariant and the fibration over S^1 will also be equivariant. M_6 admits no S^1-action.

$\mathcal{G}_2 \cong G_2$ by $\tau(\alpha) = q_1$, $\tau(t_2) = q_2^{-1}q_1$, $\tau(t_3) = q_2q_3^{-1}$

$\mathcal{G}_3 \cong G_3$ by $\tau(\alpha) = q_1^{-1}$, $\tau(t_2) = q_1^{-1}q_2$

$\mathcal{G}_4 \cong G_4$ by $\tau(\alpha) = q_2$, $\tau(t_2) = q_1q_2^{-1}$

$\mathcal{G}_5 \cong G_5$ by $\tau(\alpha) = q_1$, $\tau(t_2) = q_1^{-2}q_2$

$\mathcal{G}_6 \cong G_6$ by $\tau(\alpha) = q_1$, $\tau(\gamma) = v_1^{-1}$

For these isomorphisms the groups are reduced by Tietze transformations to have only the given generators. The isomorphism for G_5 was found by A. Strøm. It is interesting to note that the G_i are all solvable groups, see (8.3).

The next result is again due to Wolf [1,p.120].

Theorem 3. There are just 4 affine diffeomorphism classes of compact connected non-orientable flat 3-dimensional riemannian manifolds. They are represented by the manifolds R^3/Γ where Γ is one of the 4 groups $\mathcal{B}_i$ given below. Here Λ is the translation lattice, $\{a_1,a_2,a_3\}$ are its generators, $t_i = t_{a_i}$, $\Psi = \Gamma/\Gamma^*$ is the holonomy, and $\Gamma_0 = \Gamma \cap SO(3)\cdot R^3$ so that $R^3/\Gamma_0 \to R^3/\Gamma$ is the 2-sheeted orientable riemannian covering.

$\mathcal{B}_1$. $\Psi = \mathbb{Z}_2$ and Γ is generated by $\{\epsilon,t_1,t_2,t_3\}$ where $\epsilon^2 = t_1$, $\epsilon t_2\epsilon^{-1} = t_2$, $\epsilon t_3\epsilon^{-1} = t_3^{-1}$; a_1 and a_2 are orthogonal to a_3 while $\epsilon = (E,t_{a_1/2})$ with $E(a_1) = a_1$, $E(a_2) = a_2$ and $E(a_3) = -a_3$. Γ_0 is generated by $\{t_1,t_2,t_3\}$.

$\mathcal{B}_2$. $\Psi = \mathbb{Z}_2$ and Γ is generated by $\{\epsilon,t_1,t_2,t_3\}$ where $\epsilon^2 = t_1$, $\epsilon t_1\epsilon^{-1} = t_2, \epsilon t_3\epsilon^{-1} = t_1t_2t_3^{-1}$; the orthogonal projection of a_3 on the (a_1,a_2)-plane is $(a_1+a_2)/2$; $\epsilon = (E,t_{a_1/2})$ with $E(a_1) = a_1$, $E(a_2) = a_2$, $E(a_3) = a_1+a_2-a_3$. Γ_0 is generated by $\{t_1,t_2,t_3\}$.

$\mathcal{B}_3$. $\Psi = \mathbb{Z}_2 \times \mathbb{Z}_2$ and Γ is generated by $\{\varepsilon, \alpha, t_1, t_2, t_3\}$ where $\alpha^2 = t_1$, $\varepsilon^2 = t_2$, $\varepsilon\alpha\varepsilon^{-1} = t_2\alpha$, $\alpha t_2 \alpha^{-1} = t_2^{-1}$, $\alpha t_3 \alpha^{-1} = t_3^{-1}$, $\varepsilon t_1 \varepsilon^{-1} = t_1$ and $\varepsilon t_3 \varepsilon^{-1} = t_3^{-1}$; the a_i are mutually orthogonal and

$$\alpha = (A, t_{a_1/2}) \text{ with } A(a_1) = a_1,\ A(a_2) = -a_2,\ A(a_3) = -a_3,$$

$$\varepsilon = (E, t_{a_2/2}) \text{ with } E(a_1) = a_1,\ E(a_2) = a_2\ ,\ E(a_3) = -a_3.$$

Γ_0 is generated by $\{\alpha, t_1, t_2, t_3\}$.

$\mathcal{B}_4$. $\Psi = \mathbb{Z}_2 \times \mathbb{Z}_2$ and Γ is generated by $\{\varepsilon, \alpha, t_1, t_2, t_3\}$ where $\alpha^2 = t_1$, $\varepsilon^2 = t_2$, $\varepsilon\alpha\varepsilon^{-1} = t_2 t_3 \alpha$, $\alpha t_2 \alpha^{-1} = t_2^{-1}$, $\alpha t_3 \alpha^{-1} = t_3^{-1}$, $\varepsilon t_1 \varepsilon^{-1} = t_1$, $\varepsilon t_3 \varepsilon^{-1} = t_3^{-1}$; the a_i are mutually orthogonal and

$$\alpha = (A, t_{a_1/2}) \text{ with } A(a_1) = a_1,\ A(a_2) = -a_2,\ A(a_3) = -a_3,$$

$$\varepsilon = (E, t_{(a_2+a_3)/2}) \text{ with } E(a_1) = a_1,\ E(a_2) = a_2,\ E(a_3) = -a_3.$$

Γ_0 is generated by $\{\alpha, t_1, t_2, t_3\}$.

Theorem 4. The four compact connected non-orientable flat 3-dimensional riemannian manifolds are the four Klein-bottle bundles over S^1 . Let $\pi_1(K) = (x_1, x_2 | x_1^2 x_2^2)$. Then

$N_1 = \{0;(n_1,2)\} = K \times S^1$,

$N_2 = \{1;(n_1,2)\}$ is the K-bundle over S^1 with characteristic map $\psi(x_1) = x_2^{-1}$, $\psi(x_2) = x_1^{-1}$,

$N_3 = \{0;(n_3,2)\}$ is the K-bundle over S^1 with characteristic map $\psi(x_1) = x_1^{-1}$, $\psi(x_2) = x_2^{-1}$,

$N_4 = \{1;(n_3,2)\}$ is the K-bundle over S^1 with characteristic map $\psi(x_1) = x_2$, $\psi(x_2) = x_1$.

Proof. Again we let $B_i = \pi_1(N_i)$ and show that $\mathcal{B}_i \cong B_i$. Note that N_1 and N_2 admit S^1-actions while N_3 and N_4 do not.

$\mathcal{B}_1 \cong B_1$ by $\tau(\varepsilon) = v_1$, $\tau(t_2) = h$, $\tau(t_3) = v_1v_2$;

$\mathcal{B}_2 \cong B_2$ by $\tau(\varepsilon) = v_1$, $\tau(t_3) = v_1v_2$;

$\mathcal{B}_3 \cong B_3$ by $\tau(\varepsilon) = v_1v_2$, $\tau(\alpha) = v_2^{-1}$, $\tau(t_3) = h^{-1}$;

$\mathcal{B}_4 \cong B_4$ by $\tau(\varepsilon) = v_1v_2$, $\tau(\alpha) = v_1v_2v_1$.

The groups are again reduced by Tietze transformations to have only the given generators. The isomorphisms for B_3 and B_4 were found by A. Strøm. The orientable double cover is M_1 for N_1 and N_2 and M_2 for N_3 and N_4 . Clearly the B_i are also solvable groups, (8.3).

8.3. Solvable Fundamental Groups

Let G be a group and $G^{(1)} = [G,G]$ be its commutator subgroup. Define inductively $G^{(m)} = [G^{(m-1)},G^{(m-1)}]$ and call G solvable if the series terminates, i.e.

$$G \supset G^{(1)} \supset \ldots \supset G^{(m)} = 1$$

for some m . Typical example is an abelian group. A well-known example of a non-solvable group is the binary icosahedral group I^*, since $[I^*,I^*] = I^*$. The subgroups and factor groups of solvable groups are sovable and the extension of a solvable group by a solvable group is solvable. An equivalent definition is that G has a finite series of normal subgroups

$$G \supset G_1 \supset \ldots \supset G_n = 1$$

each G_i normal in G_{i-1} so that G_{i-1}/G_i is abelian for all i. If G_{i-1}/G_i is in the center of G/G_i for all i, then G is called nilpotent.

If G is the fundamental group of a Seifert manifold, then

G is solvable if and only if the planar discontinuous group $G/(h)$ is solvable. These considerations give the following result essentially due to Moser [1].

Theorem 1. The Seifert manifolds with solvable fundamental groups are:

(i) $M = \{b;(o_1,1)\}$, T^2-bundles over S^1 ; G is a nilpotent extension of $\mathbb{Z}\times\mathbb{Z}$ by $\mathbb{Z}$;

(ii) $M = \{b;(o_1,0);(2,1),(2,1),(2,1),(2,1)\}$, for $b = -2$ M is a T^2 bundle over S^1 , otherwise M is the orbit space of a free $\mathbb{Z}_2$-action on one of the manifolds of (i), G is an extension of a nilpotent group by $\mathbb{Z}_2$;

(iii) o_1, $g = 0$, $r = 3$, $\frac{1}{\alpha_1} + \frac{1}{\alpha_2} + \frac{1}{\alpha_3} \geq 1$ except for $(\alpha_1,\alpha_2,\alpha_3) = (2,3,5)$ where I^* is a direct summand of G ; for $(3,3,3)$, $(2,4,4)$ and $(2,3,6)$ M either fibers over S^1 , see (8.2.2) or it is the orbit space of one of the finite groups $\mathbb{Z}_3$, $\mathbb{Z}_4$ or $\mathbb{Z}_6$ acting freely on one of the manifolds of (i) so G is a single or double cyclic extension of a nilpotent group; for $(2,2,n)$, $(2,3,3)$ and $(2,3,4)$ G is finite, see (6.2.2);

(iv) o_1, $g = 0$, $r \leq 2$ are lens spaces or $S^2\times S^1$ so G is finite or infinite cyclic;

(v) $M = \{b;(n_2,2)\}$ are T-bundles over S^1 so G is an extension of $\mathbb{Z}\times\mathbb{Z}$ by $\mathbb{Z}$;

(vi) n_2, $g = 1$, $r \leq 1$, here $\{0;(n_2,1)\} = \mathbb{RP}^3 \# \mathbb{RP}^3$ with $G = \mathbb{Z}_2 * \mathbb{Z}_2$ which is an extension of $\mathbb{Z}$ by $\mathbb{Z}_2$ while the other manifolds have finite fundamental groups, see (6.2.2);

(vii) $M = \{b;(n_2,1);(2,1),(2,1)\}$ are orbit spaces of the free orientation preserving $\mathbb{Z}_2$ actions on manifolds of (ii) that

<u>induce the antipodal map in the orbit space of the S^1-action; G is the double extension of a nilpotent group by cyclic groups;</u>

(viii) $M = \{b;(o_2,1)\}$ <u>are</u> K-<u>bundles over</u> S^1 , <u>so</u> G <u>is</u> <u>an extension of a solvable group by</u> $\mathbb{Z}_2$;

(ix) $M = \{b;(n_1,2)\}$ <u>same as</u> (viii);

(x) $M = \{b;(n_3,2)\}$ <u>are the other two</u> K-<u>bundles over</u> S^1;

(xi) n_1, $g = 1$, $r \leq 1$ <u>are the manifolds</u> $P^2 \times S^1$ <u>and</u> N <u>so</u> G <u>is</u> $\mathbb{Z} \times \mathbb{Z}_2$ <u>or</u> $\mathbb{Z}$;

(xii) $M = \{b;(n_1,1);(2,1),(2,1)\}$ <u>are orbit spaces of the free orientation reversing</u> $\mathbb{Z}_2$ <u>actions on manifolds of</u> (ii) <u>that induce the antipodal map in the orbit space of the</u> S^1-<u>action</u>; G <u>is the double extension of a nilpotent group by cyclic groups</u>.

8.4. Finite Group Actions

If $M = \{b;(\epsilon,g);(\alpha_1,\beta_1),\ldots,(\alpha_r,\beta_r)\}$ admits an S^1-action, so $\epsilon = o_1$ or n_1 , then every finite subgroup $\mathbb{Z}_k \subset S^1$ acts on M with orbit space a Seifert manifold M' whose invariants were computed by Seifert [1,p.218]:

$$M' = \{b';(\epsilon,g);(\alpha_1',\beta_1'),\ldots,(\alpha_r',\beta_r')\}$$

where

$$b' = kb \ , \ \alpha_j' = \alpha_j/(\alpha_j,k) \ , \ \beta_j' = k\beta_j/(\alpha_j,k) \ .$$

These Seifert invariants may need normalization. The action of $\mathbb{Z}_k$ is free on M if and only if $(\alpha_j,k) = 1$ for $j = 1,\ldots,r$. Note that the homeomorphisms of the action are isotopic to the identity.

The example of M_6 in (8.2.2) shows that not every finite

group acts as a subgroup of the circle. Tollefson [2] investigates when a free $\mathbb{Z}_k$ action on a 3-manifold M embeds in an S^1-action. It is clearly necessary that a homeomorphism generating the action be homotopic to the identity. Such an action is called *proper*. Let M' be the orbit space and $\pi: M \to M'$ the orbit map. The action is called *$\mathbb{Z}$-classified* if there is a commutative diagram

$$\begin{array}{ccc} M & \longrightarrow & S^1 \\ \pi\downarrow & & \downarrow p \\ M' & \longrightarrow & S^1 \end{array}$$

where $p: S^1 \to S^1$ is the usual k-sheeted covering of the circle. In particular such maps exist if $H_1(M';\mathbb{Z})$ has no k-torsion. Two $\mathbb{Z}_k$-actions $\mu, \nu: \mathbb{Z}_k \times M \to M$ are called *weakly equivalent* if there is a group automorphism $A: \mathbb{Z}_k \to \mathbb{Z}_k$ and a homeomorphism $H: M \to M$ so that $\mu(g) = H^{-1}\nu(A(g))H$ for all $g \in \mathbb{Z}_k$. The main result of Tollefson [2] is:

Theorem 1. Let M be a closed, orientable, irreducible 3-manifold. A $\mathbb{Z}$-classified free $\mathbb{Z}_p$-action on M ($p \geq 2$ prime) is proper if and only if it is weakly equivalent to some $\mathbb{Z}_p$-action embedded in an effective S^1-action on M.

In the course of the proof it is shown that M fibers over S^1 and the $\mathbb{Z}_p$-action is equivariant with respect to the fibration.

Notice that in some cases a Seifert-manifold may cover itself, e.g. it follows from the opening remarks of this section that

$$M = \{-1;(o_1,g);(\alpha,1),(\alpha,\alpha-1)\}$$

is a proper k-sheeted covering of itself for every $k \equiv 1 \bmod \alpha$. For $g = 0$ $M = S^2 \times S^1$ but for $g > 0$ M is irreducible and a non-trivial 2-manifold bundle over S^1. Tollefson [3] proves that if M is a closed, connected 3-manifold that is a non-trivial connected sum and covers itself, then $M = \mathbb{R}P^3 \# \mathbb{R}P^3$. It is the k-fold cover of itself for every k but none of these free $\mathbb{Z}_k$-actions are proper in the above sense. If the covering action is proper, then Tollefson [3] shows that the manifold M is irreducible and if $H_1(M;\mathbb{Z})$ has no element of order k, then M fibers over S^1 .

8.5. Foliations

Let M be a smooth manifold with tangent bundle TM . A k-plane field on M is a k-dimensional subbundle σ of TM . If L is an injectively immersed, smooth submanifold of M so that $TL_x = \sigma_x \subset TM_x$ for all $x \in L$, then L is called an integral submanifold of σ . A k-plane field σ is called completely integrable if the following three equivalent conditions are satisfied:

A. M is covered by open sets U with local coordinates $x_1,\ldots,x_m$ so that the submanifolds defined by x_{k+1} = constant, $\ldots, x_m$ = constant are integral submanifolds of σ .

B. σ is smooth and through every point $x \in M$ there is an integral submanifold L of σ .

C. σ is smooth and if X and Y are vector fields on M with $X_x, Y_x \in \sigma_x$ for all $x \in M$ then the bracket $[X,Y]_x \in \sigma_x$.

An integrable k-plane field is called a foliation and the maximal connected integral submanifolds are called leaves. The leaves of a foliation partition the manifold. The following result is due independently to Lickorish, Novikov and Zieschang.

<u>Theorem 1</u>. <u>Every closed, orientable</u> 3-<u>manifold admits a codimension one foliation</u>.

The proof goes roughly as follows. The Reeb foliation on $D^2 \times S^1$ is obtained by considering a function with graph below

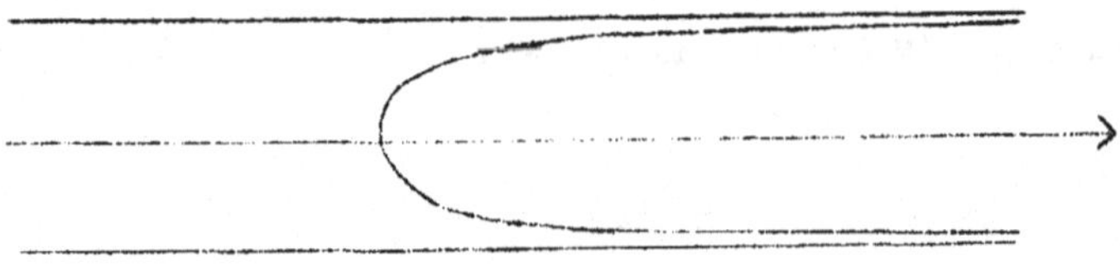

and all its translates along the x-axis. Rotate to obtain a foliation of $D^2 \times R$ and identify integral translates to obtain the Reeb foliation on $D^2 \times S^1$. It has one compact leaf, $\partial D^2 \times S^1$ and all other leaves are homeomorphic to R^2. The union of two Reeb foliations foliates S^3. Every orientable closed 3-manifold is obtained from S^3 by a finite number of (1,1)-surgeries according to Wallace. Remove the necessary number of solid tori from S^3 and alter the foliation of S^3 at the boundary tori by the procedure of "dropping off leaves"

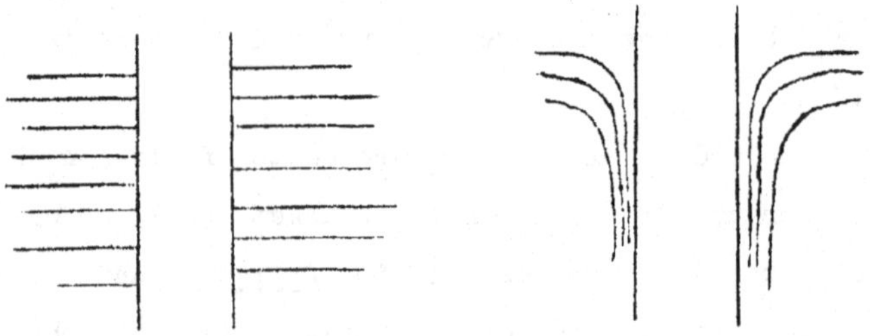

to foliate the resulting manifold. Now sew in the required copies of $D^2 \times S^1$ with Reeb foliations to obtain the manifold in question.

Wood [1] showed that non-orientable closed 3-manifolds also admit codimension one foliations. A celebrated theorem of Novikov proves that every codimension one foliation of S^3 has a compact leaf.

The <u>rank</u> of a differentiable manifold M is the maximum number of linearly independent C^2 vector fields on M which commute pairwise. If M is a closed manifold, then the rank of M is the largest integer k so that there exists a non-singular action of R^k on M with all orbits of dimension k. This action defines a foliation of M. The following was proved by Rosenberg-Roussaire-Weil [1].

<u>Theorem 2</u>. <u>Closed orientable 3-manifolds have the following rank</u>:

(i) $S^1 \times S^1 \times S^1$ <u>has rank</u> 3 ;

(ii) M <u>has rank</u> 2 <u>if and only if it is a non-trivial torus bundle over</u> S^1;

(iii) <u>all others have rank</u> 1 .

The proof is outlined in the paper as follows. If Φ is a non-singular action of R^2 on the closed, orientable manifold V, then the orbits are R^2, $R \times S^1$ or T^2. It is known that if all orbits are R^2, then V is T^3. If V has rank 2, then there must be orbits homeomorphic to $R \times S^1$ or T^2. If all orbits are homeomorphic to $R \times S^1$,
then Φ is modified to a C^0-close action Φ_1 which has a compact orbit. It is known that not every compact orbit of Φ can separate V into two connected components. One can find k compact orbits $T_1, \ldots, T_k$ which do not separate V but have the

property that for every other compact orbit T the union $T \cup T_1 \cup \dots \cup T_k$ separates V. Let W be the manifold obtained by cutting V along the T_i, $i = 1,\dots,k$. Then ∂W consists of $2k$ tori and every torus orbit in the interior separates W into connected components. By a transfinite argument it is obtained that Φ has no compact orbits in the interior of W. The crucial step is to show that $W \simeq T^2 \times [0,1]$ so V is obtained as a T^2 bundle over S^1.

An explicit action of R^2 on a T^2 bundle over S^1 is defined as follows: Let $f: T^2 \to T^2$ be the orientation preserving characteristic map of the bundle and $V = T^2 \times I/f$. As noted earlier f is isotopic to a linear map $F \in \Lambda^+(T^2) = GL^+(2,\mathbb{Z})$ and V is diffeomorphic to $T^2 \times I/F$. Since the group $GL^+(2,\mathbb{R})$ is connected there is an isotopy F_t with $F_0 = id$, $F_1 = F^{-1}$. Choose it so that $F_t = F_0$ for $t < \epsilon$ and $F_t = F_1$ for $1-\epsilon<t\leq 1$ for some small $\epsilon > 0$. Any two constant vector fields on F^2 which are linearly independent define two linearly independent commuting vector fields on T^2. For $t \in [0,1]$ let $X(t) = F_t(1,0)$ and $Y(t) = F_t(0,1)$. Then $X(t)$ and $Y(t)$ are two linearly independent vector fields on $T^2 \times t$. Moreover, $dF_1(X(1)) = (0,1) = X(0)$ and $dF_1(Y(1)) = (0,1) = Y(0)$, hence $X(t)$ and $Y(t)$ define two linearly independent vector fields on V.

It is interesting to note that if V has no compact orbits, then $F = \begin{pmatrix} 1 & a \\ 0 & 1 \end{pmatrix}$, so V is the Seifert manifold $\{-a;(o_1,1)\}$.

8.6. Flows

A C^r <u>flow</u> on a C^r manifold M is a C^r action $\mu: M \times R \to M$ of the additive reals on M. Such actions arise naturally from the integration of a C^r vector field on M. Conversely, differentiation of a C^{r+1} flow gives rise to a C^r vector field on M.

The following is an example of a flow on $S^3 = \{(z_1,z_2) \in C^2 \mid z_1\bar{z}_1+z_2\bar{z}_2 = 1\}$. Let (p,q) be relatively prime integers and define

$$\mu(z_1,z_2,t) = (z_1e^{2\pi ipt}, z_2e^{2\pi iqt}) .$$

This is clearly the R action obtained from lifting the corresponding S^1 action to the universal cover of S^1. For $p = q = 1$ this is called the <u>Hopf flow</u> on S^3. These flows have only closed orbits. The following recent result of Epstein [1] proves that if all orbits are closed on a 3-manifold, then this is the most general situation.

<u>Theorem 1</u>. *Let $\mu: M \times R \to M$ be a C^r action $(1 \leq r \leq \infty)$ of the additive group of real numbers on M, with every orbit a circle. Let M be a compact 3-manifold possibly with boundary. Then there is a C^r action $\mu': M \times S^1 \to M$ with the same orbits as μ.*

If non-compact orbits are present, then the structure of flows is still unknown. The following result is due to Seifert [2]. Let C be the vector field of Clifford-parallel vectors whose integral curves, the Clifford circles, give the Hopf flow and let $\tilde{C}$ be a continuous vector field on S^3 which differs sufficiently little from C, that is, the angle between a vector of C and that of $\tilde{C}$ is at every point of S^3 smaller than a sufficiently small α.

<u>Theorem 2</u>. *A continuous vector field on the 3-sphere which differs sufficiently little from the field of Clifford-parallels and which sends through every point exactly one integral curve has at least one closed integral curve.*

The question posed by Seifert [2] whether this is true for all flows on S^3 is still open and is now referred to as the Seifert Conjecture.*

---

* Added in proof: Paul Schweitzer has obtained a counterexample to this conjecture.

References

M.F. Atiyah and I.M. Singer

1. The index of elliptic operators III, Ann. of Math. 87 (1968), 546-604.

E. Brieskorn

1. Über die Auflösung gewisser Singularitäten von holomorphen Abbildungen, Math. Ann. 166 (1966), 76-102.

C. Chevalley

1. Séminaire 1-2, Paris 1956/58.

P.E. Conner and E.E. Floyd

1. Maps of odd period, Ann. of Math. 84 (1966), 132-156.
2. Differentialbe periodic maps, Springer Verlag, 1964.

P.E. Conner and F. Raymond

1. Injective operations of the toral groups, Topology 10 (1971), 283-296.

D.B.A. Epstein

1. Periodic flows on three-manifolds, Ann. of Math. 95 (1972), 66-82.

W. Fulton

1. Algebraic curves, Benjamin, New York, 1969.

R.C. Gunning

1. Lectures on complex analytic varieties, Princeton University Press, 1970.

H. Hironaka

1. Resolution of singularities of an algebraic variety over a field of characteristic zero, Ann. of Math.79 (1964),109-326.

F. Hirzebruch

1. Differentialbe manifolds and quadratic forms, revised by W.D. Neumann, Marcel Dekker Inc., New York, 1972.
2. Topological methods in Algebraic Geometry, Springer Verlag,1966.

H. Holmann

1. Seifertsche Faserräume, Math. Ann. 157 (1964), 138-166.

H. Hopf

1. Zum Clifford-Kleinschen Raumproblem, Math. Ann. 95 (1926), 313-319.

W. Jaco

1. Surfaces embedded in $M^2 \times S^1$, Can.J.Math. 22 (1970),553-568.

K. Jänich

1. Differenzierbare G-Mannigfaltigkeiten, Springer Verlag, Lecture notes no.59, 1968.

J. Milnor

1. Singular points of complex hypersurfaces, Princeton University Press, 1968.
2. Groups which act on S^n without fixed points, Amer.J.Math.79 (1957), 623-630.

J. Milnor and P. Orlik

1. Isolated singularities defined by weighted homogeneous polynomials, Topology 9 (1970), 385-393.

D. Montgomery and L. Zippin

1. Topological transformation groups, Interscience, New York,1955.

L. Moser

1. Elementary surgery along torus knots and solvable fundamental groups of closed 3-manifolds, Thesis, University of Wisconsin, 1970.

D. Mumford

1. The topology of normal singularities of an algebraic surface and a criterion for simplicity, Publ. Math. No.9. IHES, Paris, 1961.
2. Geometric Invariant Theory, Academic Press. New York, 1965.

W.D. Neumann

1. S^1-actions and the α-invariant of their involutions, Bonner Mathematische Schriften 44, 1970.

L. Neuwirth

1. A topological classification of certain 3-manifolds, Bull. Amer. Math. Soc. 69 (1963), 372-375.

P. Orlik

1. On the extensions of the infinite cyclic group by a 2-manifold group, Ill.J.Math. 12 (1968), 479-482.

P. Orlik and F. Raymond

1. Actions of SO(2) on 3-manifolds, in Proceedings of the Conference on Transformation Groups, Springer Verlag, 1968, 297-318.
2. On 3-manifolds with local SO(2) action, Quart.J.Math. Oxford 20 (1969), 143-160.

P. Orlik, E. Vogt and H. Zieschang

1. Zur Topologie gefaserter dreidimensionaler Mannigfaltigkeiten, Topology 6 (1967), 49-64.

P. Orlik and P. Wagreich

1. Isolated singularities of algebraic surfaces with C* action. Ann. of Math. 93 (1971), 205-228.
2. Singularities of algebraic surfaces with C* action, Math. Ann. 193 (1971), 121-135.

E. Ossa

1. Cobordismustheorie von fixpunktfreien und semifreien S^1-Mannigfaltigkeiten, Thesis, Bonn 1969.

E. Prill

1. Über lineare Faserräume und schwach negative holomorphe Geradenbündel, Math. Zeitschr. 105 (1968), 313-326.

R. von Randow

1. Zur Topologie von dreidimensionalen Baummannigfaltigkeiten, Bonner Mathematische Schriften 14, 1962.

F. Raymond

1. Classification of the actions of the circle on 3-manifolds, Trans. Amer. Math. Soc. 131 (1968), 51-78.

H. Rosenberg, R. Roussaire and D. Weil
1. A classification of closed orientable 3-manifolds of rank two, Ann. of Math. 91 (1970), 449-464.

M. Rosenlicht
1. On quotient varieties and the affine embedding of certain homogeneous spaces, Trans. Amer. Math. Soc. 101 (1961), 211-231.

H. Seifert
1. Topologie dreidimensionaler gefaserter Räume, Acta math. 60 (1933), 147-238.
2. Closed integral curves in 3-space and isotopic 2-dimensional deformations, Proc. Amer. Math. Soc. 1 (1950), 287-302.

H. Seifert and W. Threlfall
1. Topologische Untersuchungen der Diskontinuitätsbereiche endlicher Bewegungsgruppen des dreidimensionalen sphärischen Raumes I, Math. Ann. 104 (1931), 1-70; II, Math. Ann. 107 (1933), 543-596.

J. Stallings
1. On fibering certain 3-manifolds, in Topology of 3-manifolds, Prentice Hall, 1962, 95-103.

J. Tollefson
1. 3-manifolds fibering over S^1 with non-unique connected fiber, Proc. Amer. Math. Soc. 21 (1969), 79-80.
2. Imbedding free cyclic group actions in circle group actions, Proc. Amer. Math. Soc 26 (1970), 671-673.
3. On 3-manifolds that cover themselves, Mich. Math. J. 16 (1969), 103-109.

F. Waldhausen
1. Eine Klasse von 3-dimensionalen Mannigfaltigkeiten I, Invent. math. 3(1967), 308-333; II, Invent. math. 4 (1967), 87-117.
2. On irreducible 3-manifolds which are sufficiently large, Ann. of Math. 87 (1968), 56-88.
3. Gruppen mit Zentrum und 3-dimensionale Mannigfaltigkeiten, Topology 6 (1967), 505-517.

J. Wolf

1. Spaces of constant curvature, McGraw-Hill, 1967.

J. Wood

1. Foliations on 3-manifolds, Ann. of Math. 89 (1969), 336-358.

H. Zieschang

1. Über Automorphismen ebener diskontinuierlicher Gruppen, Math. Ann. 166 (1966), 148-167.

Vol. 178: Th. Bröcker und T. tom Dieck, Kobordismentheorie. XVI, 191 Seiten. 1970. DM 18,–

Vol. 179: Seminaire Bourbaki – vol. 1968/69. Exposés 347-363. IV. 295 pages. 1971. DM 22,–

Vol. 180: Séminaire Bourbaki – vol. 1969/70. Exposés 364-381. IV, 310 pages. 1971. DM 22,–

Vol. 181: F. DeMeyer and E. Ingraham, Separable Algebras over Commutative Rings. V, 157 pages. 1971. DM 16,–

Vol. 182: L. D. Baumert. Cyclic Difference Sets. VI, 166 pages. 1971. DM 16,–

Vol. 183: Analytic Theory of Differential Equations. Edited by P. F. Hsieh and A. W. J. Stoddart. VI, 225 pages. 1971. DM 20,–

Vol. 184: Symposium on Several Complex Variables, Park City, Utah, 1970. Edited by R. M. Brooks. V, 234 pages. 1971. DM 20,–

Vol. 185: Several Complex Variables II, Maryland 1970. Edited by J. Horváth. III, 287 pages. 1971. DM 24,–

Vol. 186: Recent Trends in Graph Theory. Edited by M. Capobianco/ J. B. Frechen/M. Krolik. VI, 219 pages. 1971. DM 18,–

Vol. 187: H. S. Shapiro, Topics in Approximation Theory. VIII, 275 pages. 1971. DM 22,–

Vol. 188: Symposium on Semantics of Algorithmic Languages. Edited by E. Engeler. VI, 372 pages. 1971. DM 26,–

Vol. 189: A. Weil, Dirichlet Series and Automorphic Forms. V, 164 pages. 1971. DM 16,–

Vol. 190: Martingales. A Report on a Meeting at Oberwolfach, May 17-23, 1970. Edited by H. Dinges. V, 75 pages. 1971. DM 16,–

Vol. 191: Séminaire de Probabilités V. Edited by P. A. Meyer. IV, 372 pages. 1971. DM 26,–

Vol. 192: Proceedings of Liverpool Singularities – Symposium I. Edited by C. T. C. Wall. V, 319 pages. 1971. DM 24,–

Vol. 193: Symposium on the Theory of Numerical Analysis. Edited by J. Ll. Morris. VI, 152 pages. 1971. DM 16,–

Vol. 194: M. Berger, P. Gauduchon et E. Mazet. Le Spectre d'une Variété Riemannienne. VII, 251 pages. 1971. DM 22,–

Vol. 195: Reports of the Midwest Category Seminar V. Edited by J.W. Gray and S. Mac Lane. III, 255 pages. 1971. DM 22,–

Vol. 196: H-spaces – Neuchâtel (Suisse)- Août 1970. Edited by F. Sigrist, V, 156 pages. 1971. DM 16,–

Vol. 197: Manifolds – Amsterdam 1970. Edited by N. H. Kuiper. V, 231 pages. 1971. DM 20,–

Vol. 198: M. Hervé, Analytic and Plurisubharmonic Functions in Finite and Infinite Dimensional Spaces. VI, 90 pages. 1971. DM 16,–

Vol. 199: Ch. J. Mozzochi, On the Pointwise Convergence of Fourier Series. VII, 87 pages. 1971. DM 16,–

Vol. 200: U. Neri, Singular Integrals. VII, 272 pages. 1971. DM 22,–

Vol. 201: J. H. van Lint, Coding Theory. VII, 136 pages. 1971. DM 16,–

Vol. 202: J. Benedetto, Harmonic Analysis on Totally Disconnected Sets. VIII, 261 pages. 1971. DM 22,–

Vol. 203: D. Knutson, Algebraic Spaces. VI, 261 pages. 1971. DM 22,–

Vol. 204: A. Zygmund, Intégrales Singulières. IV, 53 pages. 1971. DM 16,–

Vol. 205: Séminaire Pierre Lelong (Analyse) Année 1970. VI, 243 pages. 1971. DM 20,–

Vol. 206: Symposium on Differential Equations and Dynamical Systems. Edited by D. Chillingworth. XI, 173 pages. 1971. DM 16,–

Vol. 207: L. Bernstein, The Jacobi-Perron Algorithm – Its Theory and Application. IV, 161 pages. 1971. DM 16,–

Vol. 208: A. Grothendieck and J. P. Murre, The Tame Fundamental Group of a Formal Neighbourhood of a Divisor with Normal Crossings on a Scheme. VIII, 133 pages. 1971. DM 16,–

Vol. 209: Proceedings of Liverpool Singularities Symposium II. Edited by C. T. C. Wall. V, 280 pages. 1971. DM 22,–

Vol. 210: M. Eichler, Projective Varieties and Modular Forms. III, 118 pages. 1971. DM 16,–

Vol. 211: Théorie des Matroïdes. Edité par C. P. Bruter. III, 108 pages. 1971. DM 16,–

Vol. 212: B. Scarpellini, Proof Theory and Intuitionistic Systems. VII, 291 pages. 1971. DM 24,–

Vol. 213: H. Hogbe-Nlend, Théorie des Bornologies et Applications. V, 168 pages. 1971. DM 18,–

Vol. 214: M. Smorodinsky, Ergodic Theory, Entropy. V, 64 pages. 1971. DM 16,–

Vol. 215: P. Antonelli, D. Burghelea and P. J. Kahn, The Concordance-Homotopy Groups of Geometric Automorphism Groups. X, 140 pages. 1971. DM 16,–

Vol. 216: H. Maaß, Siegel's Modular Forms and Dirichlet Series. VII, 328 pages. 1971. DM 20,–

Vol. 217: T. J. Jech, Lectures in Set Theory with Particular Emphasis on the Method of Forcing. V, 137 pages. 1971. DM 16,–

Vol. 218: C. P. Schnorr, Zufälligkeit und Wahrscheinlichkeit. IV, 212 Seiten 1971. DM 20,–

Vol. 219: N. L. Alling and N. Greenleaf, Foundations of the Theory of Klein Surfaces. IX, 117 pages. 1971. DM 16,–

Vol. 220: W. A. Coppel, Disconjugacy. V, 148 pages. 1971. DM 16,–

Vol. 221: P. Gabriel und F. Ulmer, Lokal präsentierbare Kategorien. V, 200 Seiten. 1971. DM 18,–

Vol. 222: C. Meghea, Compactification des Espaces Harmoniques. III, 108 pages. 1971. DM 16,–

Vol. 223: U. Felgner, Models of ZF-Set Theory. VI, 173 pages. 1971. DM 16,–

Vol. 224: Revètements Etales et Groupe Fondamental. (SGA 1). Dirigé par A. Grothendieck XXII, 447 pages. 1971. DM 30,–

Vol. 225: Théorie des Intersections et Théorème de Riemann-Roch. (SGA 6). Dirigé par P. Berthelot, A. Grothendieck et L. Illusie. XII, 700 pages. 1971. DM 40,–

Vol. 226: Seminar on Potential Theory, II. Edited by H. Bauer. IV, 170 pages. 1971. DM 18,–

Vol. 227: H. L. Montgomery, Topics in Multiplicative Number Theory. IX, 178 pages. 1971. DM 18,–

Vol. 228: Conference on Applications of Numerical Analysis. Edited by J. Ll. Morris. X, 358 pages. 1971. DM 26,–

Vol. 229: J. Väisälä, Lectures on n-Dimensional Quasiconformal Mappings. XIV, 144 pages. 1971. DM 16,–

Vol. 230: L. Waelbroeck, Topological Vector Spaces and Algebras. VII, 158 pages. 1971. DM 16,–

Vol. 231: H. Reiter, L^1-Algebras and Segal Algebras. XI, 113 pages. 1971. DM 16,–

Vol. 232: T. H. Ganelius, Tauberian Remainder Theorems. VI, 75 pages. 1971. DM 16,–

Vol. 233: C. P. Tsokos and W. J. Padgett. Random Integral Equations with Applications to Stochastic Systems. VII, 174 pages. 1971. DM 18,–

Vol. 234: A. Andreotti and W. Stoll. Analytic and Algebraic Dependence of Meromorphic Functions. III, 390 pages. 1971. DM 26,–

Vol. 235: Global Differentiable Dynamics. Edited by O. Hájek, A. J. Lohwater, and R. McCann. X, 140 pages. 1971. DM 16,–

Vol. 236: M. Barr, P. A. Grillet, and D. H. van Osdol. Exact Categories and Categories of Sheaves. VII, 239 pages. 1971, DM 20,–

Vol. 237: B. Stenström. Rings and Modules of Quotients. VII, 136 pages. 1971. DM 16,–

Vol. 238: Der kanonische Modul eines Cohen-Macaulay-Rings. Herausgegeben von Jürgen Herzog und Ernst Kunz. VI, 103 Seiten. 1971. DM 16,–

Vol. 239: L. Illusie, Complexe Cotangent et Déformations I. XV, 355 pages. 1971. DM 26,–

Vol. 240: A. Kerber, Representations of Permutation Groups I. VII, 192 pages. 1971. DM 18,–

Vol. 241: S. Kaneyuki, Homogeneous Bounded Domains and Siegel Domains. V, 89 pages. 1971. DM 16,–

Vol. 242: R. R. Coifman et G. Weiss, Analyse Harmonique Non-Commutative sur Certains Espaces. V, 160 pages. 1971. DM 16,–

Vol. 243: Japan-United States Seminar on Ordinary Differential and Functional Equations. Edited by M. Urabe. VIII, 332 pages. 1971. DM 26,–

Vol. 244: Séminaire Bourbaki – vol. 1970/71. Exposés 382–399. IV, 356 pages. 1971. DM 26,–

Vol. 245: D. E. Cohen, Groups of Cohomological Dimension One. V, 99 pages. 1972. DM 16,–